Plastics Projects and Techniques

McKNIGHT & McKNIGHT PUBLISHING COMPANY • BLOOMINGTON, ILLINOIS

Plastics Projects and Techniques

ALVIN R. LAPPIN, Ed. D.
Associate Professor of Industrial Arts
San Jose State College
San Jose, California

FIRST EDITION

Lithographed in U. S. A.

Library of Congress Catalog Card Number: 64:22304

Introduction

Since the early nineteen forties many books have included basic information on cutting, heating, forming, and finishing plastics. This book has been prepared to supplement these publications with a variety of suggested projects. It also gives instruction on some of the more unusual ways of working plastics.

The equipment included in Section Two has been patterned after that used in industry. Along with the projects, the equipment has been built and tested at the high school level. Projects are presented in order from simple to more complex problems, and it is recommended that the beginning students select pieces in this sequence. This practice should help to develop a background for original design problems or redesigned plastics projects. Colors of plastics are suggested in the lists of materials needed. While these have been found to be pleasing, other selections or combinations of color should be used where appropriate.

The annotated list of references is provided for those who desire basic information on working plastics. Persons interested in continuing their study to gain greater depth in the use of these materials and processes will also find helpful materials listed. The author sincerely hopes that this book will prove adequate and acceptable to those who are seeking knowledge of design and fabricating techniques with plastics.

Acknowledgments

The author wishes to express his appreciation to those students who helped to develop the projects included in this book and particularly:

To Mr. Ronald Smith for information and illustrations concerning fiber glass,

To Mr. Ping Yee, and, especially, Mr. A. Charles Tyler for their work in helping to illustrate this book.

Alvin R. Lappin

Contents

SECTION ONE

Projects

For best results, first look at the photographs of the project and the drawing. Next, make certain that you have the necessary knowledge and equipment for working the particular plastic. Then list the sizes or amount of materials needed. Analyze the problem by studying the drawing and materials to prepare a plan of procedure before starting the project.

PROJECT 1

Pin Tray

Materials Required

1/8 Clear Acrylic
Novelty Lacquer (Cryst-L-Craze)

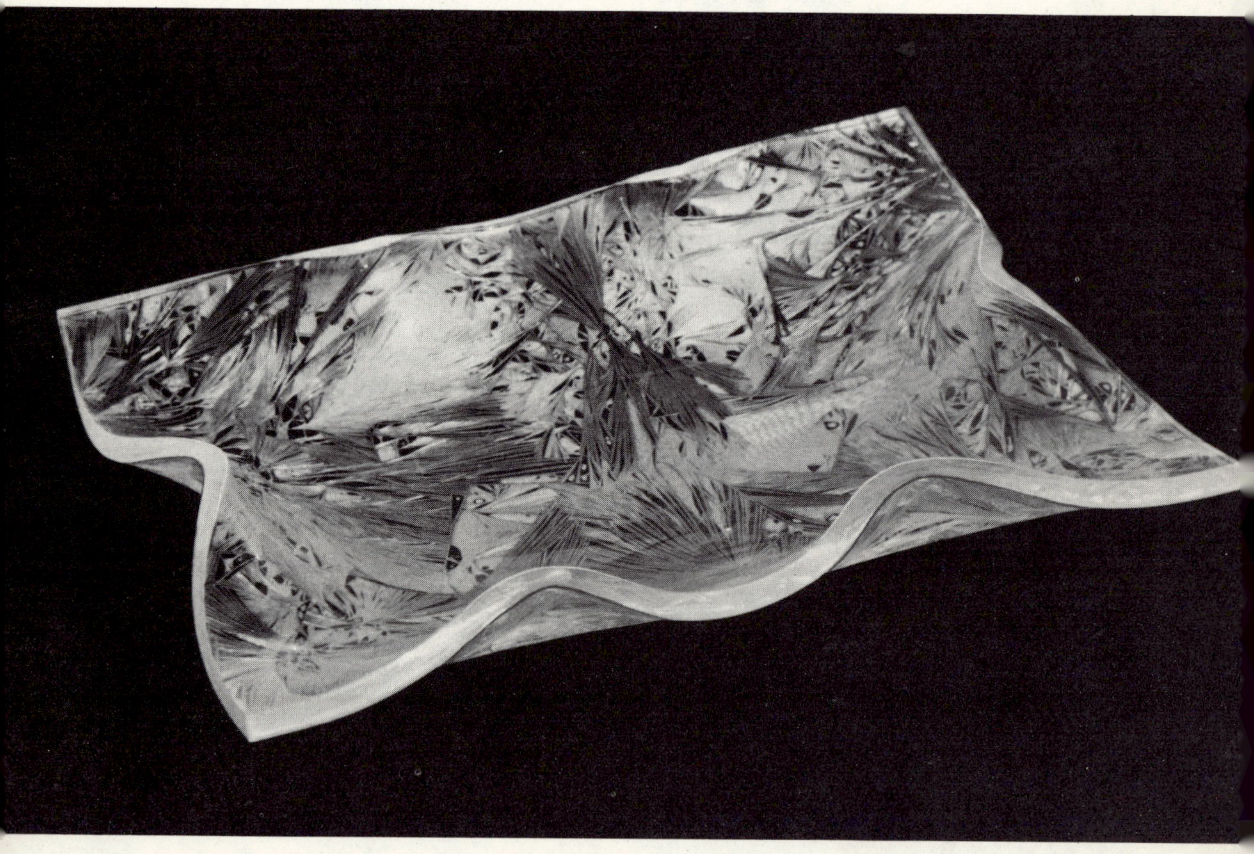

This project appeared earlier in
Industrial Arts and Vocational Education.

PIN TRAY

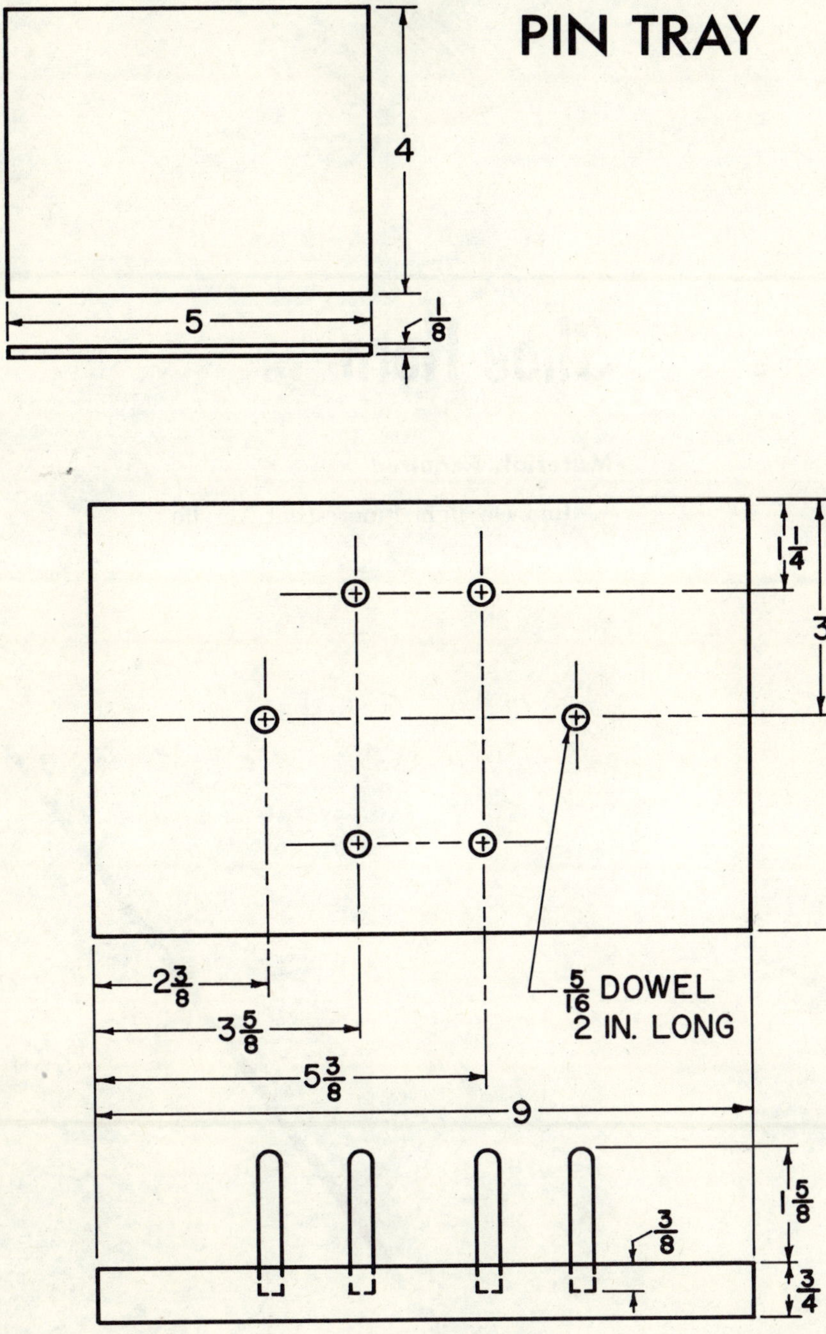

FORMING JIG

PROJECT 2

Shoe Horn

Materials Required

⅛ Red Daylight Fluorescent Acrylic

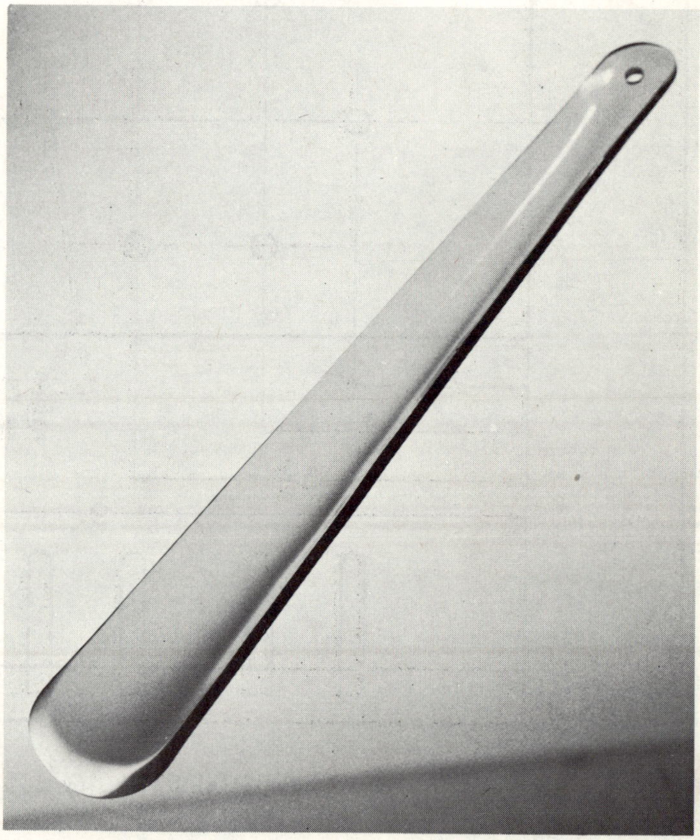

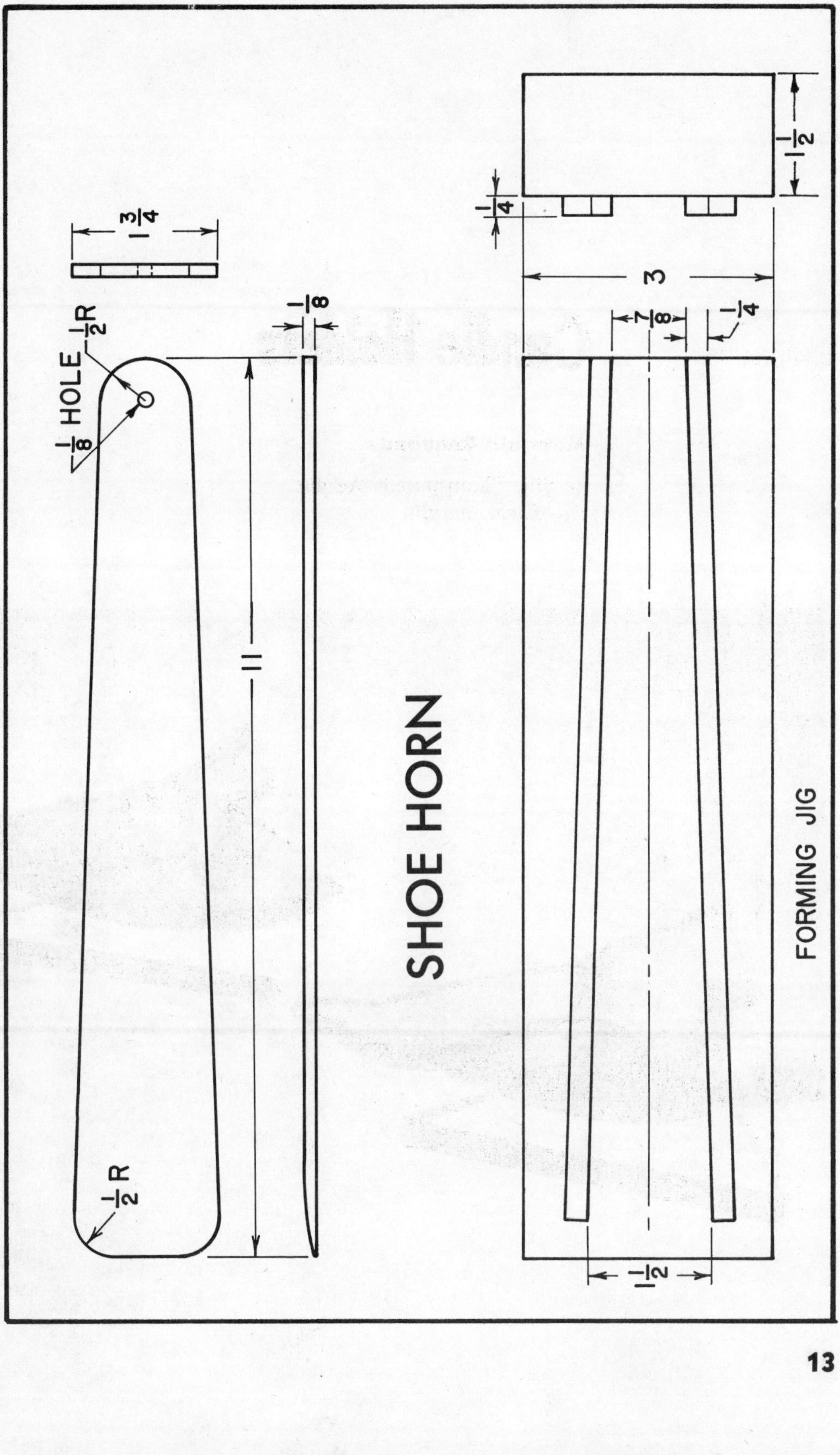
SHOE HORN
1/8 HOLE
1/2 R
1 3/4
1/8
11
1/2 R
FORMING JIG
1 1/2
1/4
3
7/8
1/4
1 1/2

PROJECT 3

Candle Holders

Materials Required

$\frac{1}{8}$ Blue Transparent Acrylic
$\frac{3}{16}$ Clear Acrylic

CANDLE HOLDERS

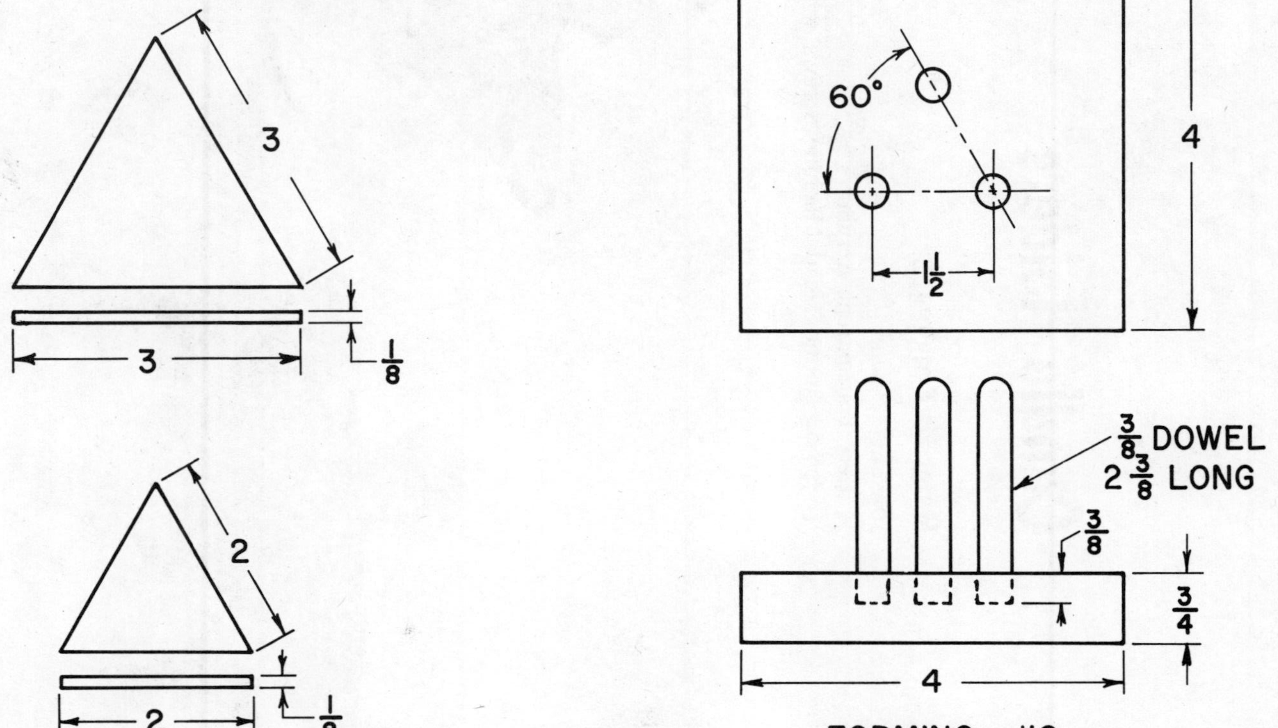

FORMING JIG

PROJECT 4

Candle Holders

Materials Required

16-Gauge Pewter
1/8 Red Transparent Acrylic
1/2 #8-32 Round Head Machine Screws with Nuts

By John Wines

CANDLE HOLDERS

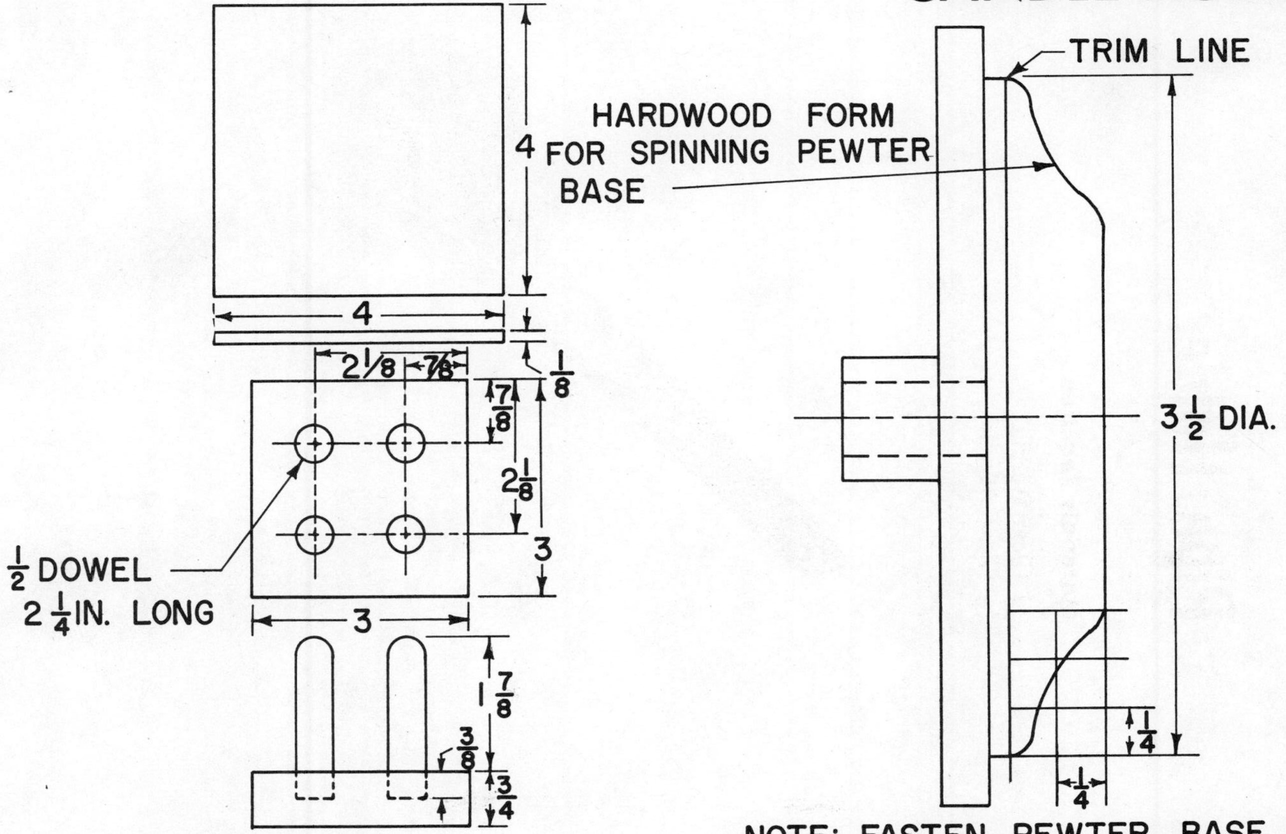

FORMING JIG

NOTE: FASTEN PEWTER BASE TO PLASTIC HOLDER WITH R.H. MACHINE SCREW.

PROJECT 5

Pipe Holder

Materials Required

¼ Clear Acrylic
⅛ Clear Acrylic

PIPE HOLDER

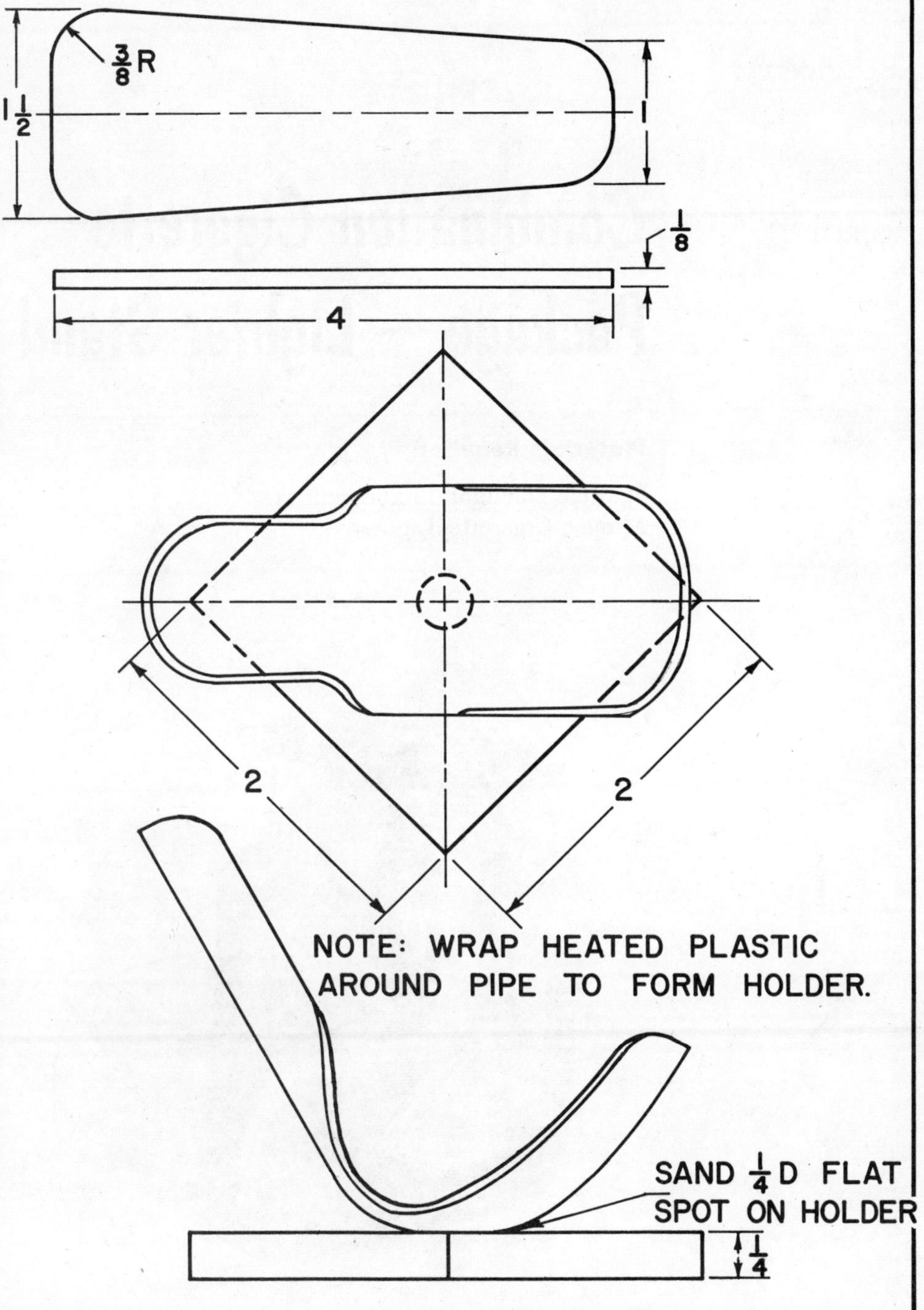
3/8 R
1 1/2
1
1/8
4
2
2
NOTE: WRAP HEATED PLASTIC AROUND PIPE TO FORM HOLDER.
SAND 1/4 D FLAT SPOT ON HOLDER
1/4

PROJECT 6

Combination Cigarette Package — Lighter Stand

Materials Required

⅛ Red Daylight Fluorescent Acrylic
Midget Cigarette Lighter

COMBINATION CIGARETTE PACKAGE — LIGHTER STAND

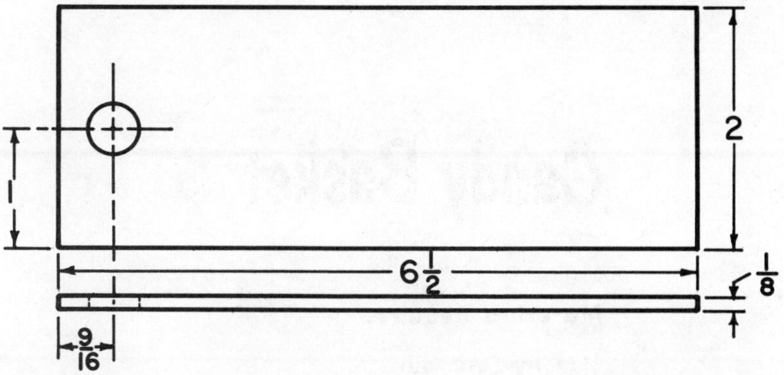

NOTE: DRILL HOLE TO FIT LIGHTER

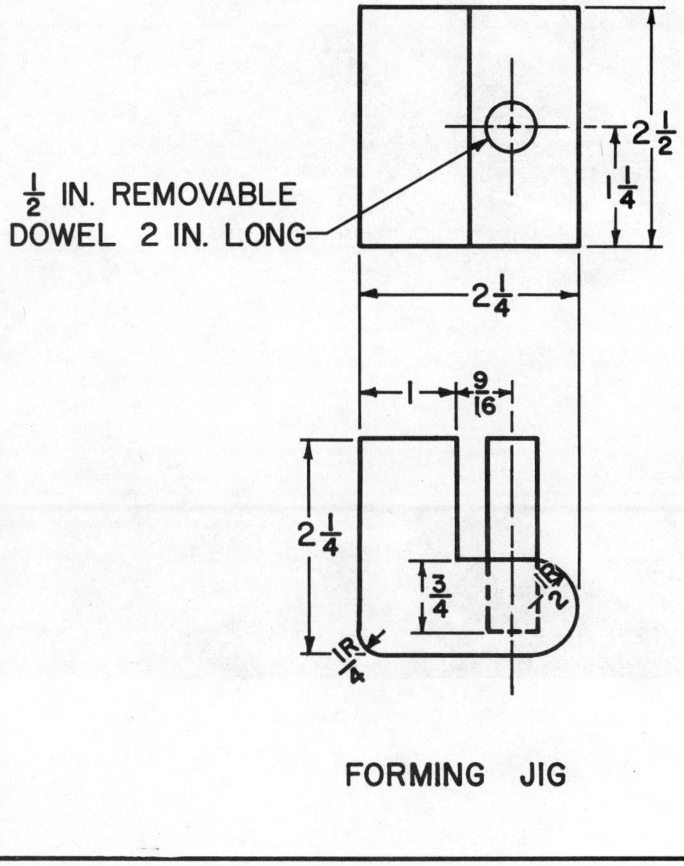

FORMING JIG

PROJECT 7

Candy Basket

Materials Required

1/8 Clear Acrylic

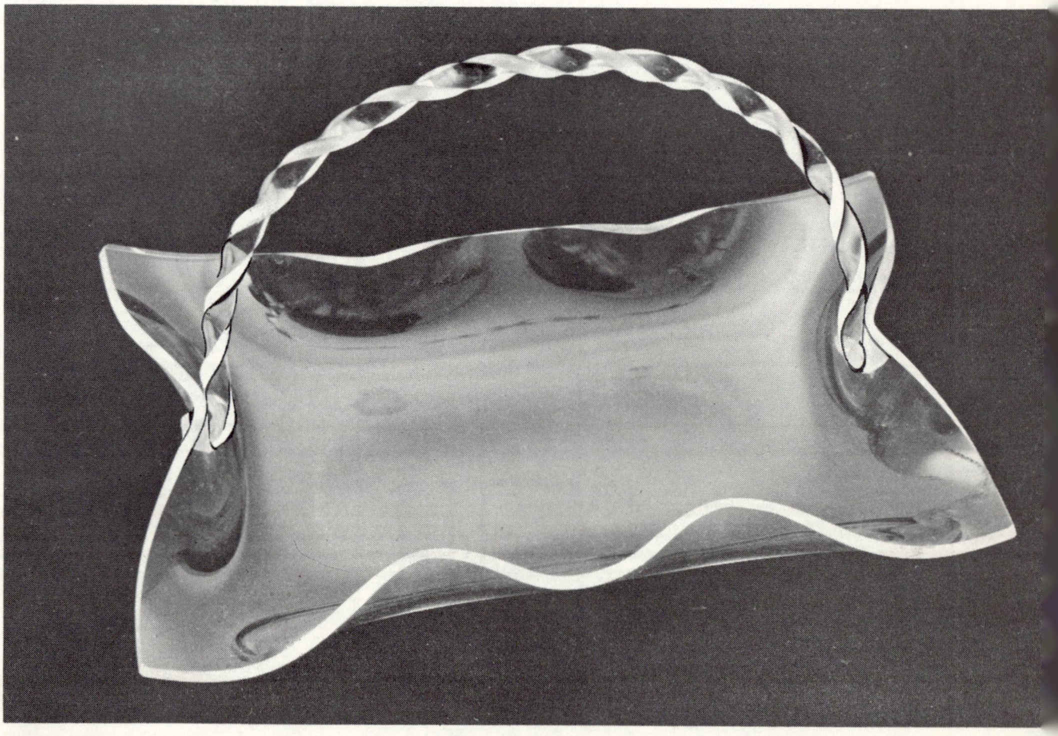

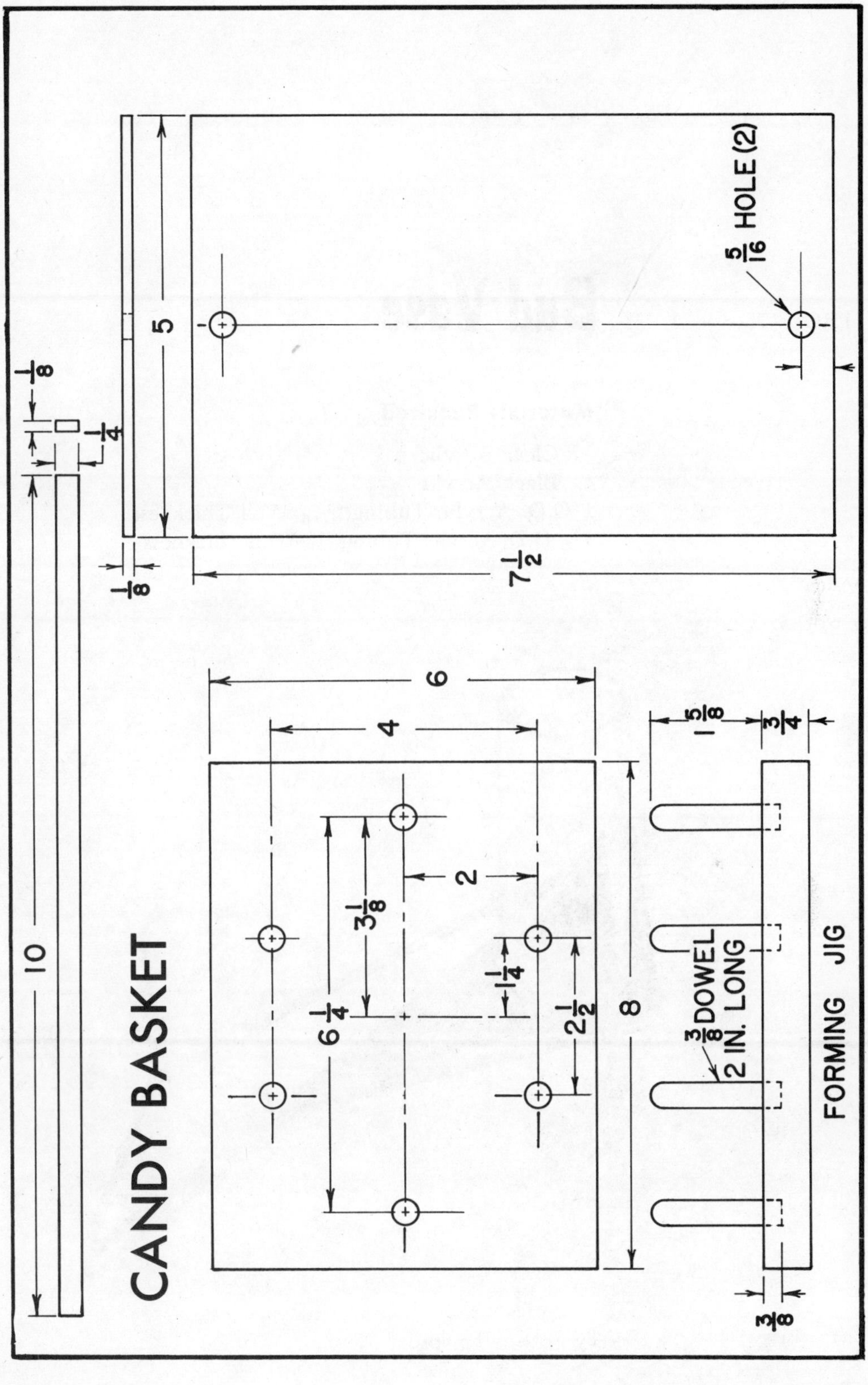
CANDY BASKET
10
1/8
1/4
5
1/8
7 1/2
5/16 HOLE (2)
6
4
2
3 1/8
6 1/4
1 1/4
2 1/2
8
3/8 DOWEL
2 IN. LONG
1 5/8
3/4
3/8
FORMING JIG

PROJECT 8

Bud Vase

Materials Required

¼ Clear Acrylic
¼ Black Acrylic
1 O.D. Acrylic Tubing, $\frac{1}{16}$ Wall Thickness
1½ O.D. Acrylic Tubing, ⅛ Wall Thickness

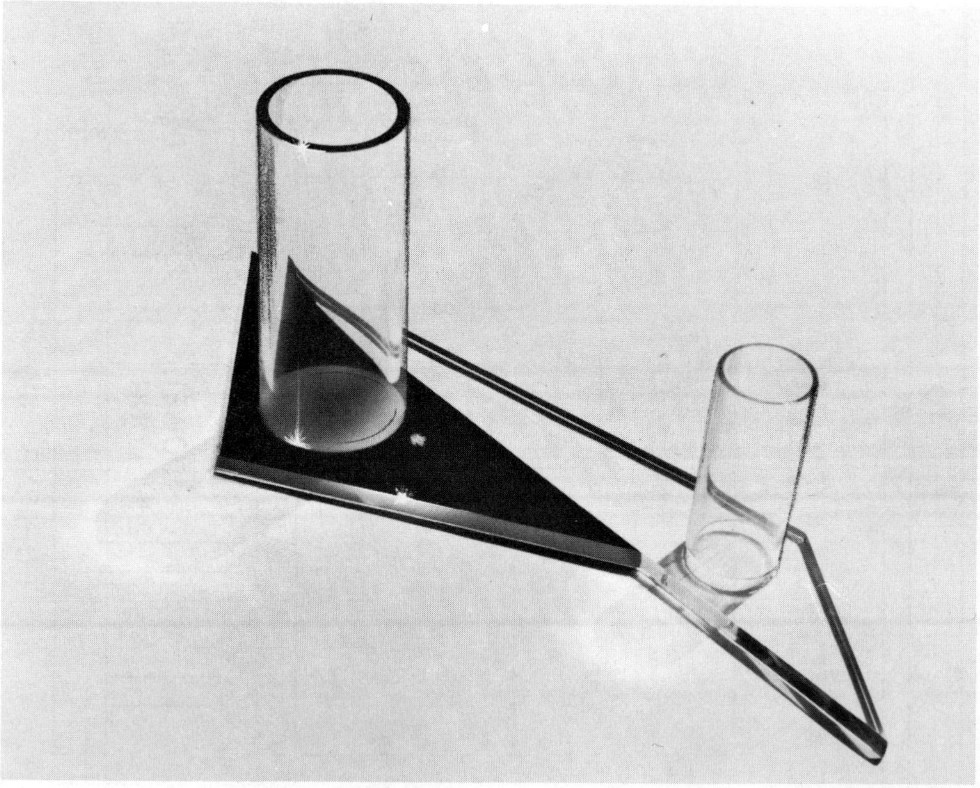

By Robert Thompson

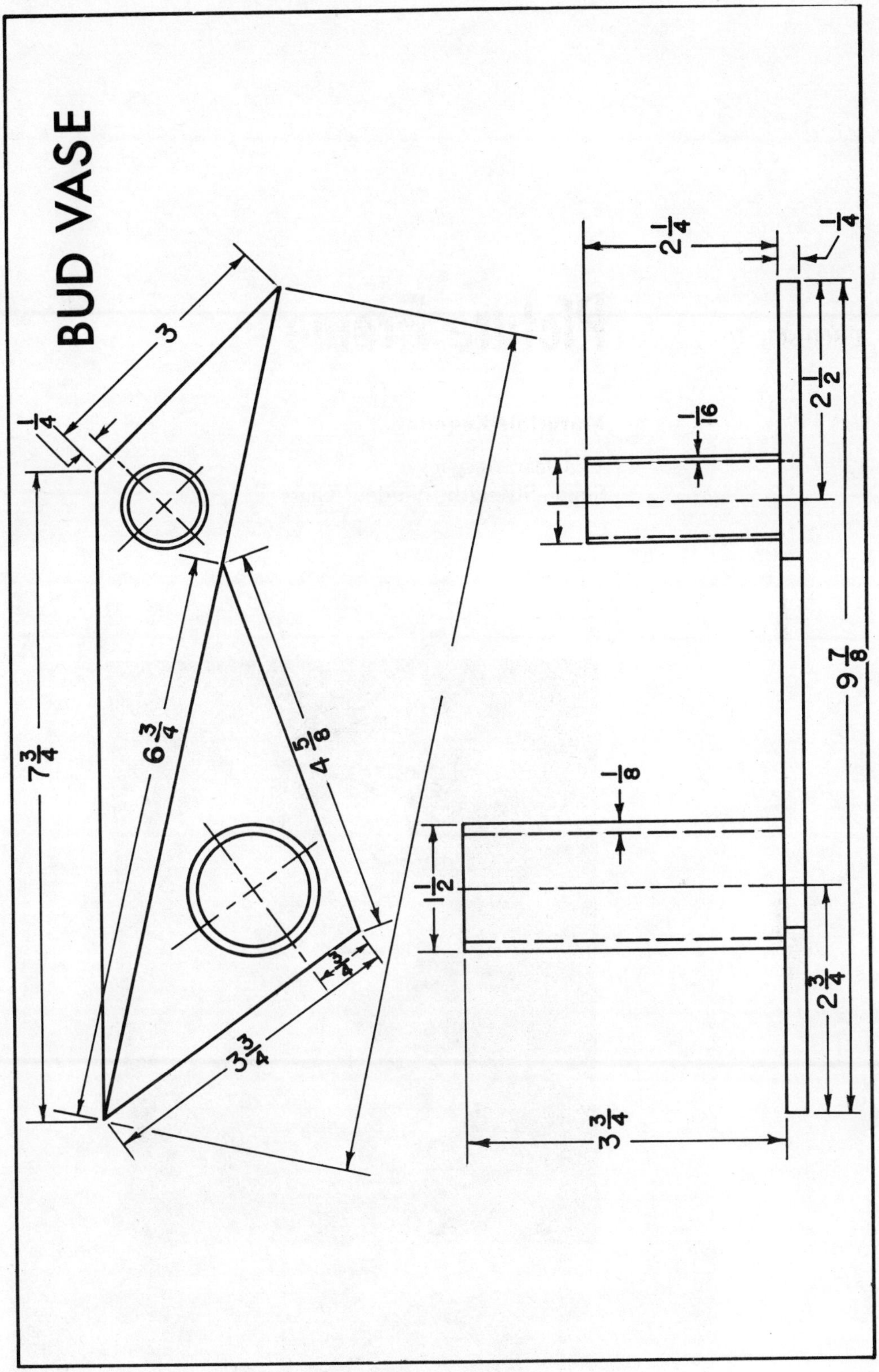
BUD VASE
3
1/4
7 3/4
6 3/4
4 5/8
3/4
3 3/4
2 1/4
1/4
1/16
2 1/2
9 7/8
1/8
1 1/2
2 3/4
3 3/4

PROJECT 9

Picture Frame

Materials Required

¼ Clear Acrylic
Single-Strength Window Glass

PICTURE FRAME

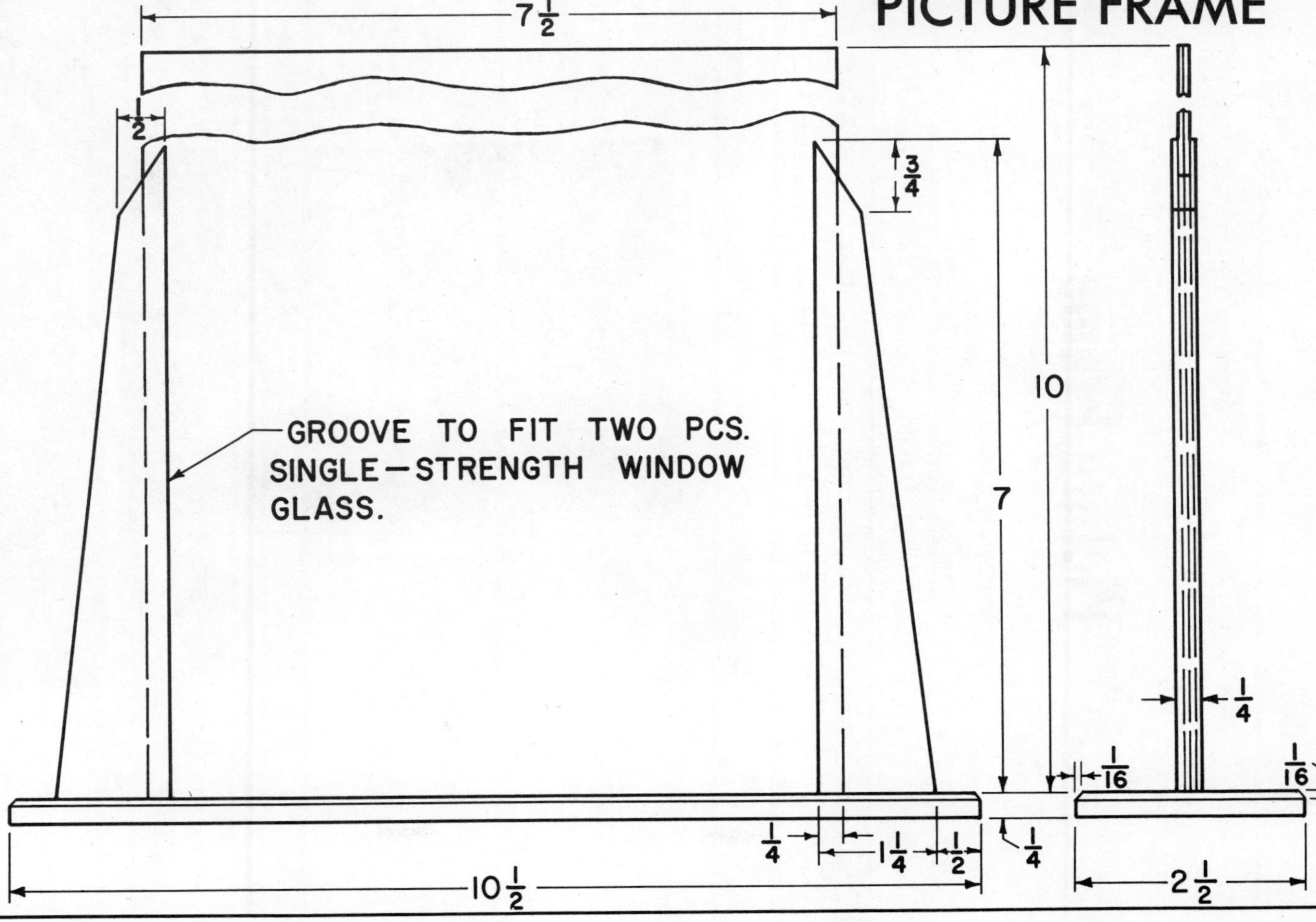

PROJECT 10

Picture Frame

Materials Required

¼ Clear Acrylic
1/16 Clear Acrylic
⅜ Dia. Acrylic Rod
16-Gauge Brass Escutcheon Pins

Refer to Technique 3 for information on strip heating.

By Mervin Yue

PICTURE FRAME

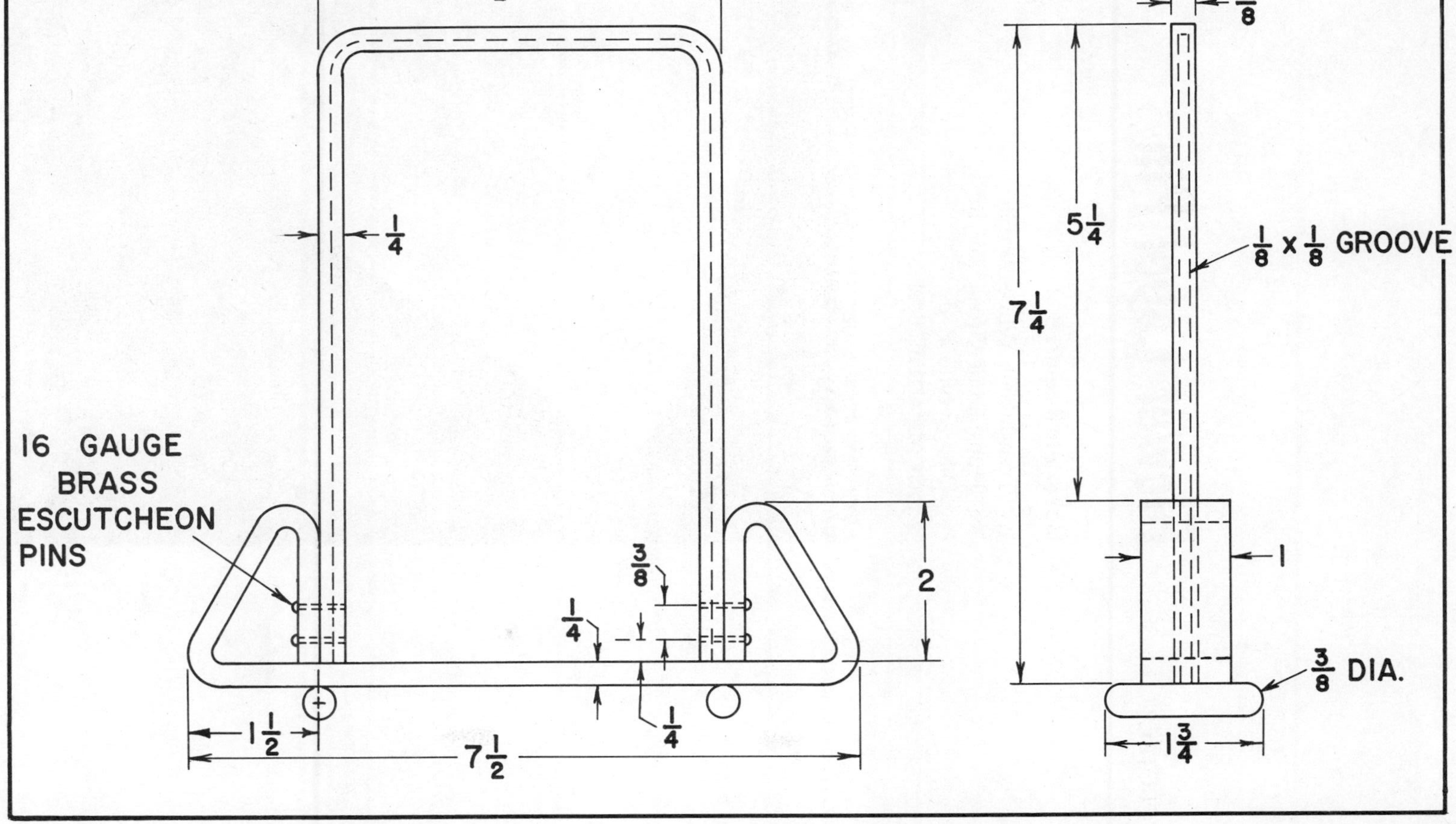

PROJECT 11

Flower Lapel Pin

Materials Required

$\frac{1}{16}$ Translucent White Acrylic
$\frac{1}{16}$ Transparent Green Acrylic
$\frac{3}{16}$ Dia. Acrylic Rod
$\frac{3}{4}$-inch Metal Pin Back

FLOWER LAPEL PIN

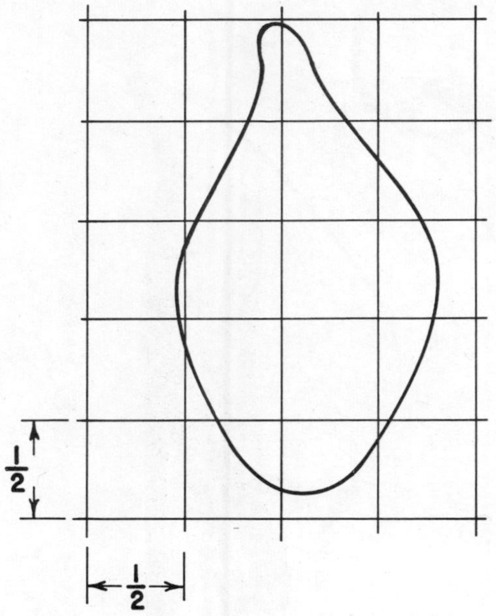

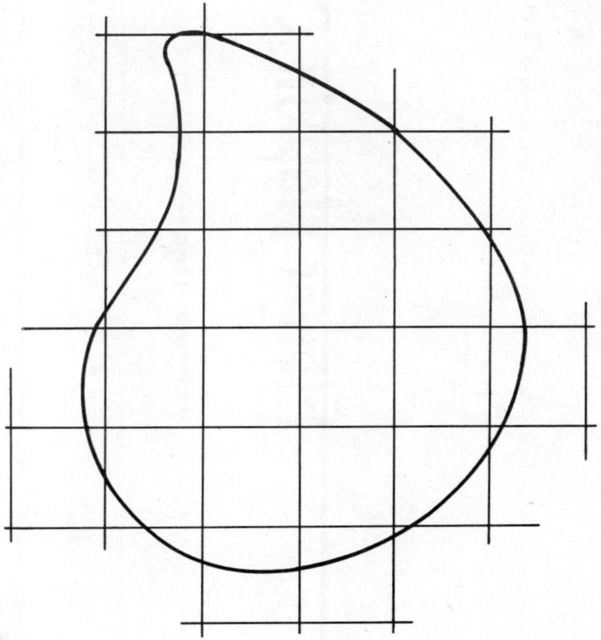

PROJECT 12

Flower Napkin Clips

Materials Required

1/8 Clear Acrylic
3/16 Dia. Acrylic Rod

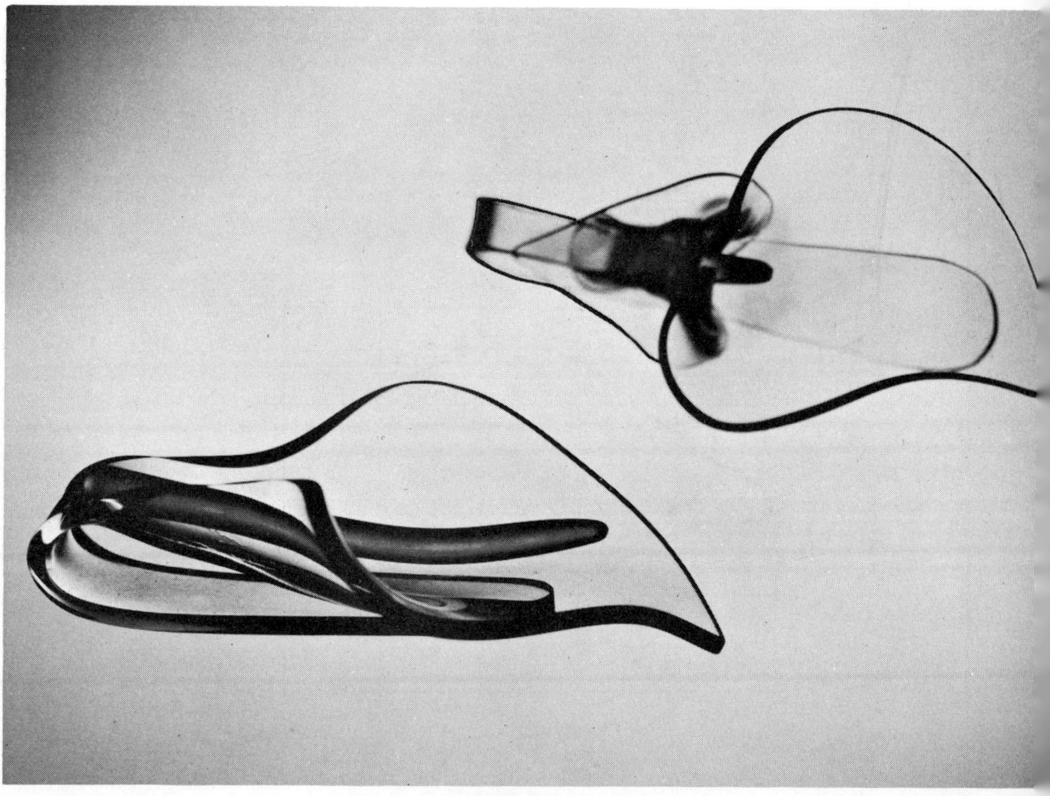

FLOWER NAPKIN CLIPS

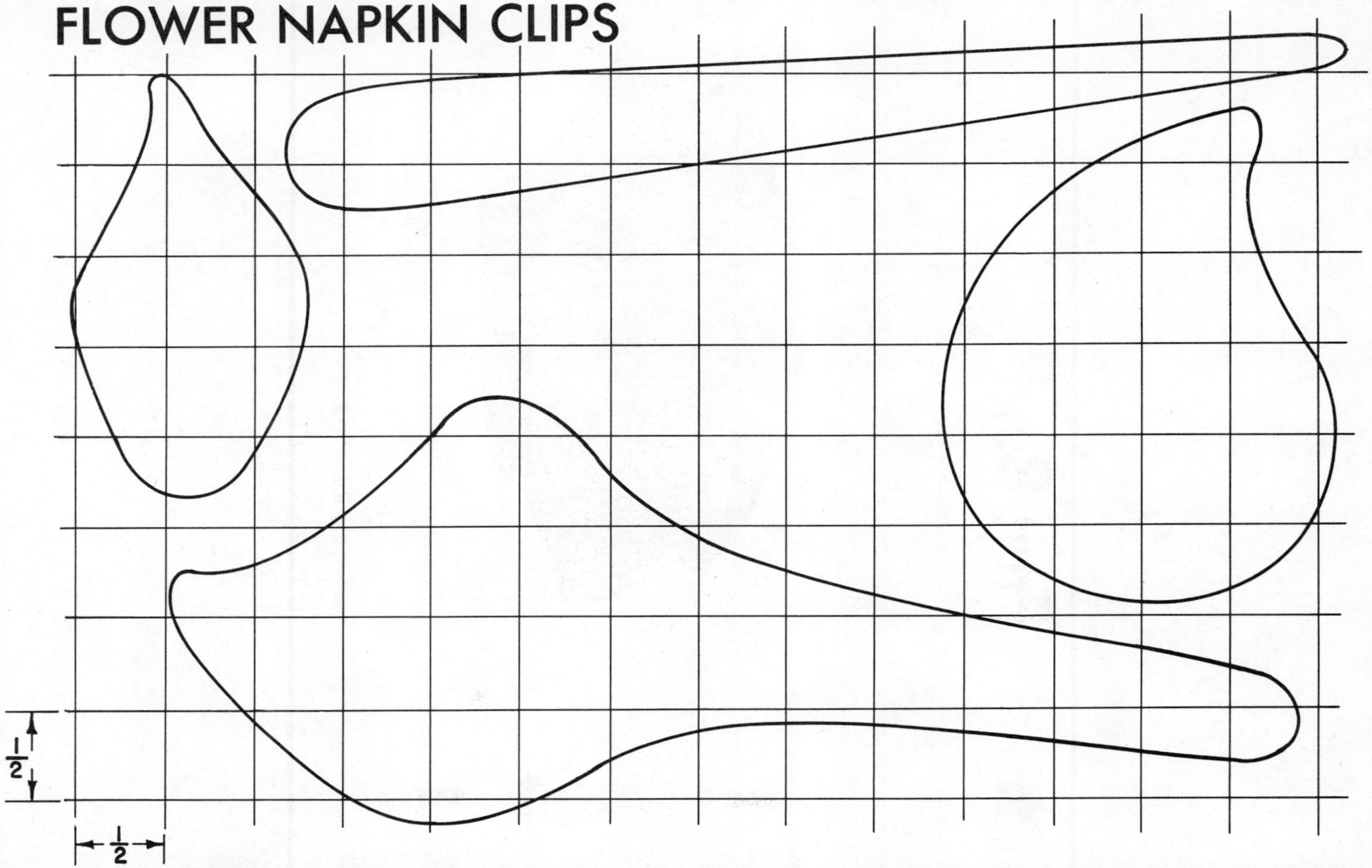

PROJECT 13

Fruit Bowl

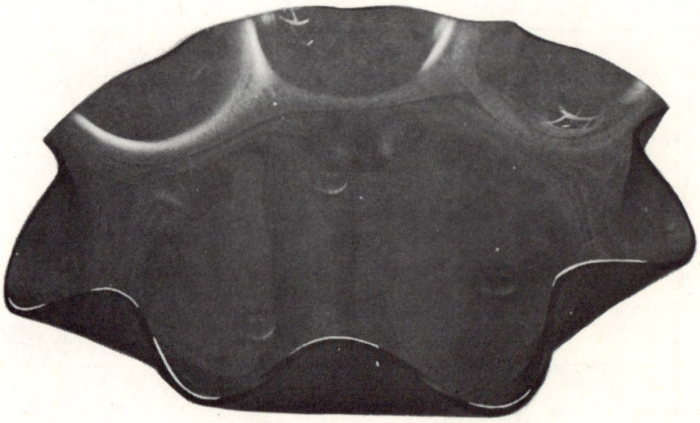

Materials Required

⅛ Transparent Green Acrylic
⅝ Dia. Acrylic Balls

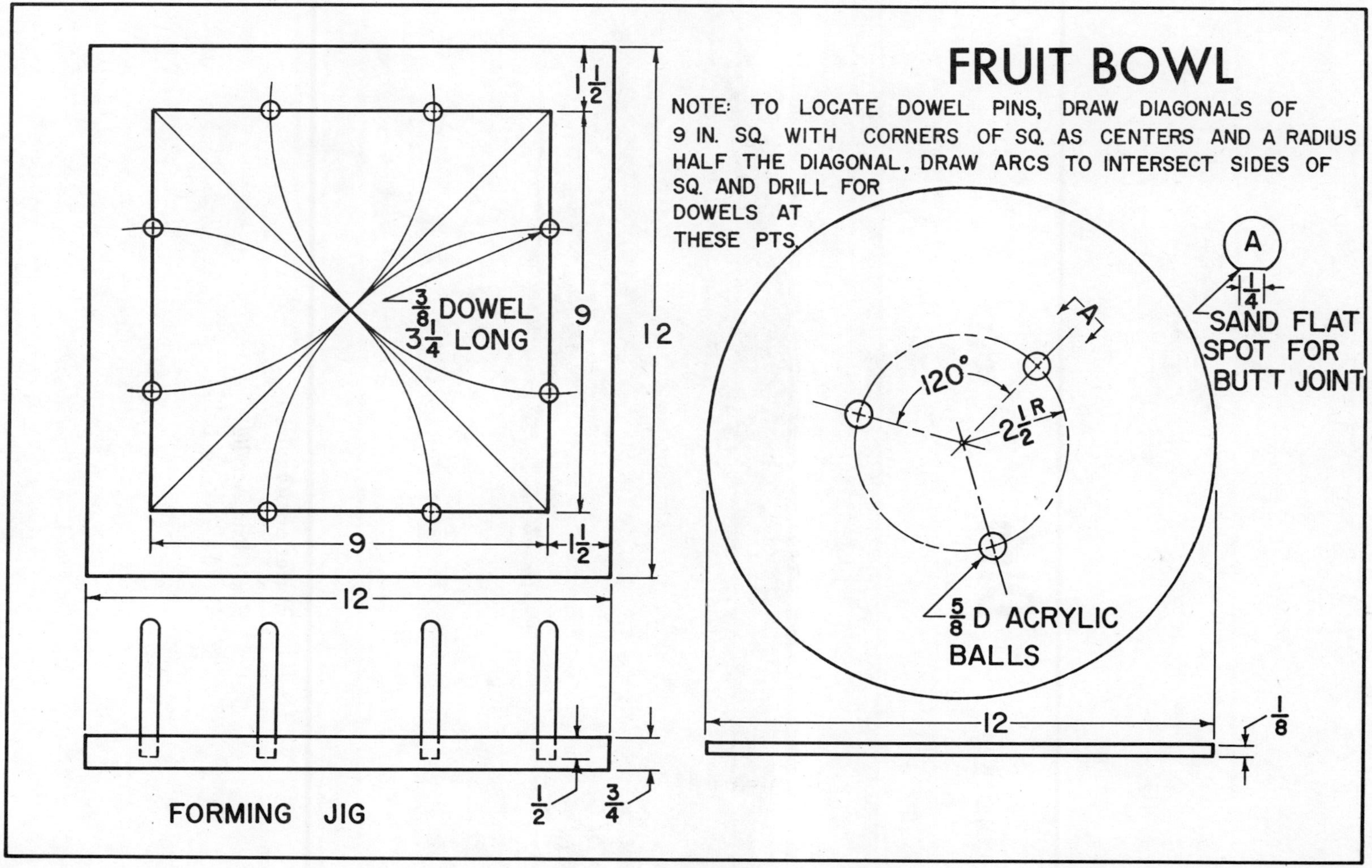
FRUIT BOWL
NOTE: TO LOCATE DOWEL PINS, DRAW DIAGONALS OF 9 IN. SQ. WITH CORNERS OF SQ. AS CENTERS AND A RADIUS HALF THE DIAGONAL, DRAW ARCS TO INTERSECT SIDES OF SQ. AND DRILL FOR DOWELS AT THESE PTS.
3/8 DOWEL 3 1/4 LONG
9
12
1 1/2
FORMING JIG
1/2
3/4
A
1/4
SAND FLAT SPOT FOR BUTT JOINT
120°
2 1/2 R
5/8 D ACRYLIC BALLS
1/8

PROJECT 14

Dish

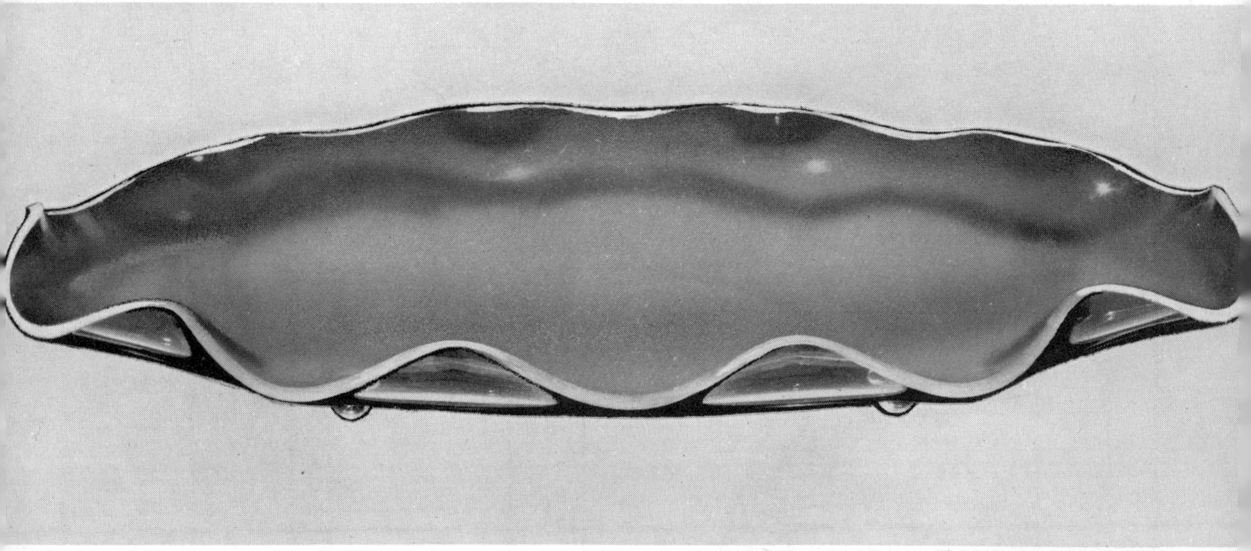

By Meyer Scharlack

Materials Required

3/16 Clear Acrylic
5/8 Dia. Acrylic Balls

DISH

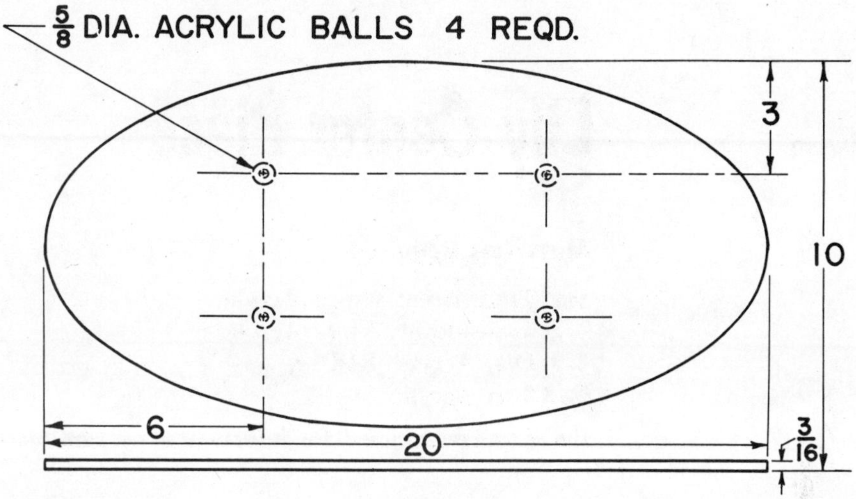

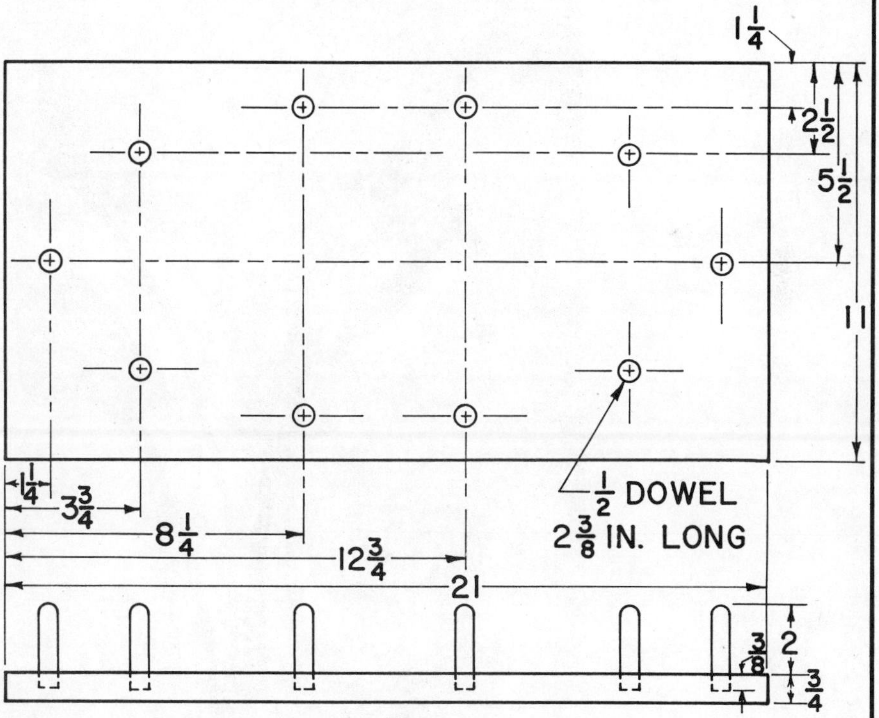

FORMING JIG

PROJECT 15

Lily Center Piece

Materials Required

$\frac{1}{16}$ Transparent Green Acrylic
$\frac{1}{16}$ Translucent White Acrylic
$\frac{3}{16}$ Dia. Acrylic Rod
$\frac{3}{8}$ Clear Acrylic

Refer to Technique 1 for information on sandblasting.

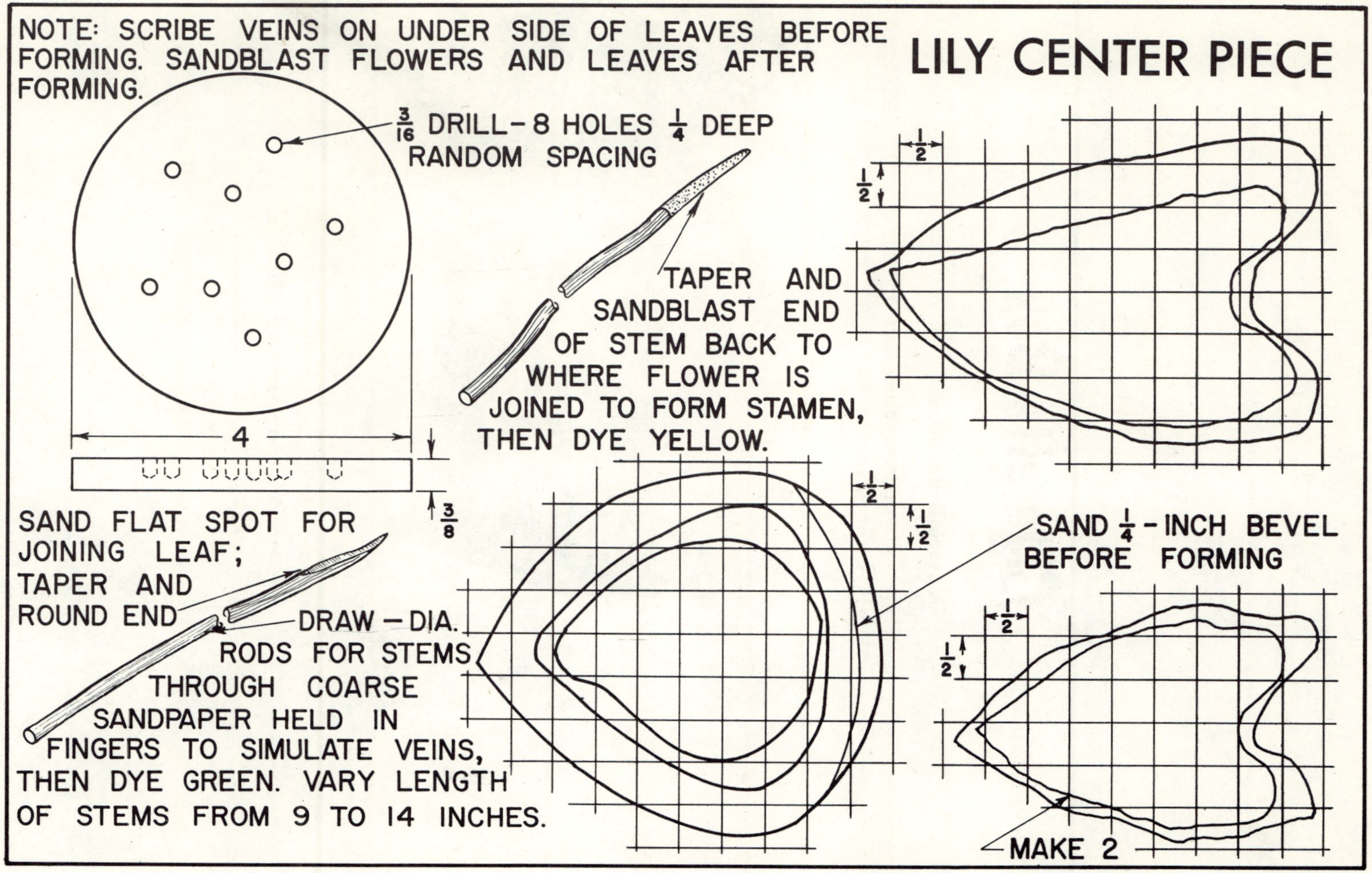
NOTE: SCRIBE VEINS ON UNDER SIDE OF LEAVES BEFORE FORMING. SANDBLAST FLOWERS AND LEAVES AFTER FORMING.
LILY CENTER PIECE
3/16 DRILL – 8 HOLES 1/4 DEEP RANDOM SPACING
4
3/8
TAPER AND SANDBLAST END OF STEM BACK TO WHERE FLOWER IS JOINED TO FORM STAMEN, THEN DYE YELLOW.
1/2
1/2
SAND FLAT SPOT FOR JOINING LEAF; TAPER AND ROUND END
DRAW – DIA. RODS FOR STEMS THROUGH COARSE SANDPAPER HELD IN FINGERS TO SIMULATE VEINS, THEN DYE GREEN. VARY LENGTH OF STEMS FROM 9 TO 14 INCHES.
SAND 1/4 - INCH BEVEL BEFORE FORMING
MAKE 2

PROJECT 16

Cigarette Box

Materials Required

3/16 Transparent Blue Acrylic
1/8 Clear Acrylic
1/16 Clear Acrylic

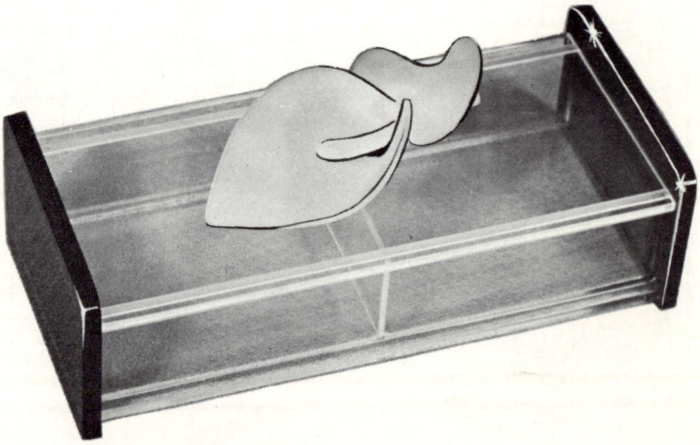

This project appeared earlier in
Industrial Arts and Vocational Education.

CIGARETTE BOX

NOTE: A PLASTIC FLOWER, PROJECT 11, IS CEMENTED TO THE BOX LID FOR A HANDLE.

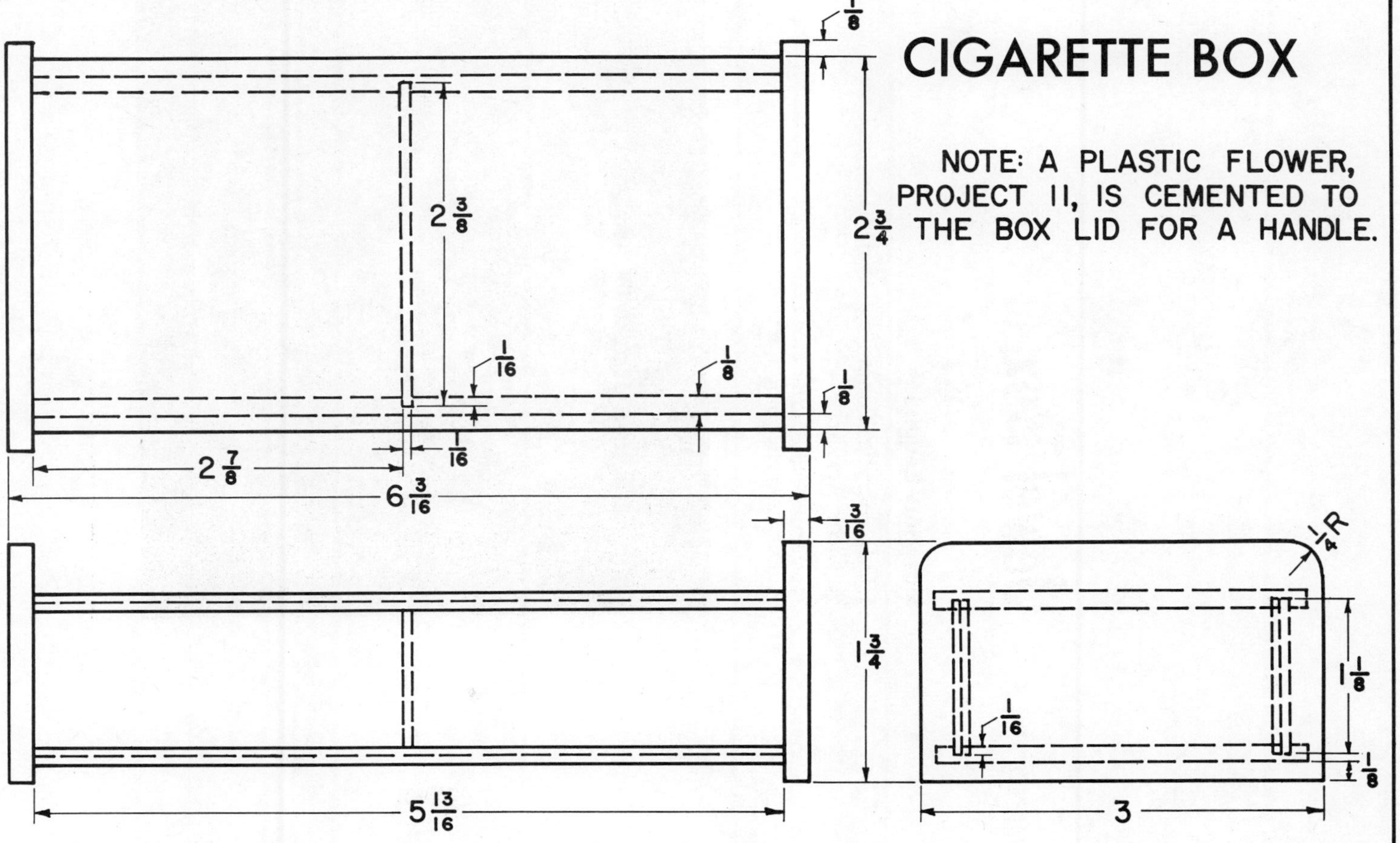

PROJECT 17

Jewel Box

Materials Required

4 O.D. Acrylic Tubing, ⅛ Wall Thickness
⅛ Clear Acrylic
½ Dia. Acrylic Rod
⅝ Dia. Acrylic Ball

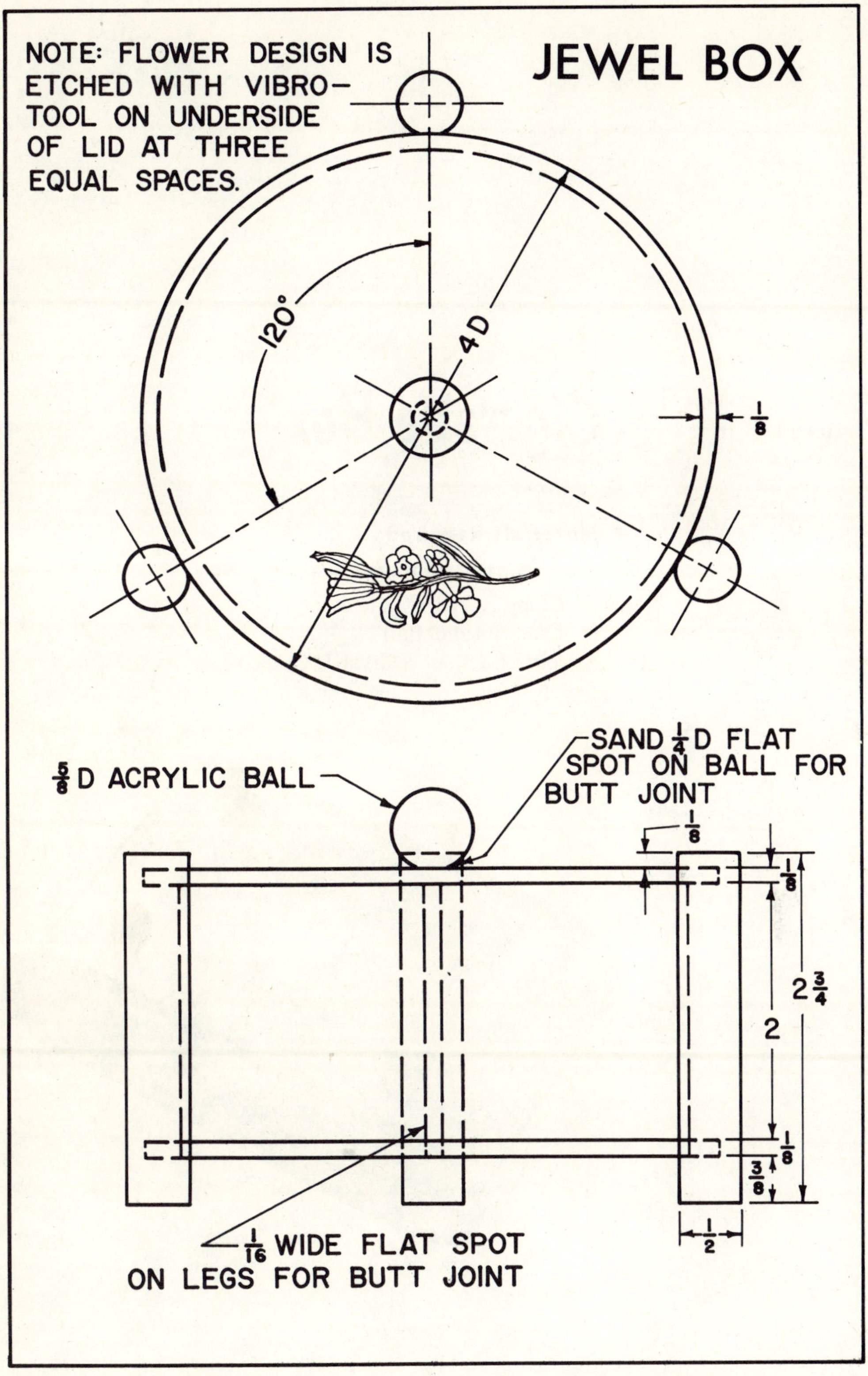
NOTE: FLOWER DESIGN IS ETCHED WITH VIBRO-TOOL ON UNDERSIDE OF LID AT THREE EQUAL SPACES.
JEWEL BOX
120°
4 D
1/8
5/8 D ACRYLIC BALL
SAND 1/4 D FLAT SPOT ON BALL FOR BUTT JOINT
1/8
1/8
2 3/4
2
1/8
3/8
1/2
1/16 WIDE FLAT SPOT ON LEGS FOR BUTT JOINT

PROJECT 18

Cigarette Box

Materials Required

$^{3}/_{16}$ Translucent Red Acrylic
$^{1}/_{8}$ Clear Acrylic
$^{5}/_{8}$ Dia. Acrylic Ball
Novelty Lacquer (Cryst-L-Craze)

CIGARETTE BOX

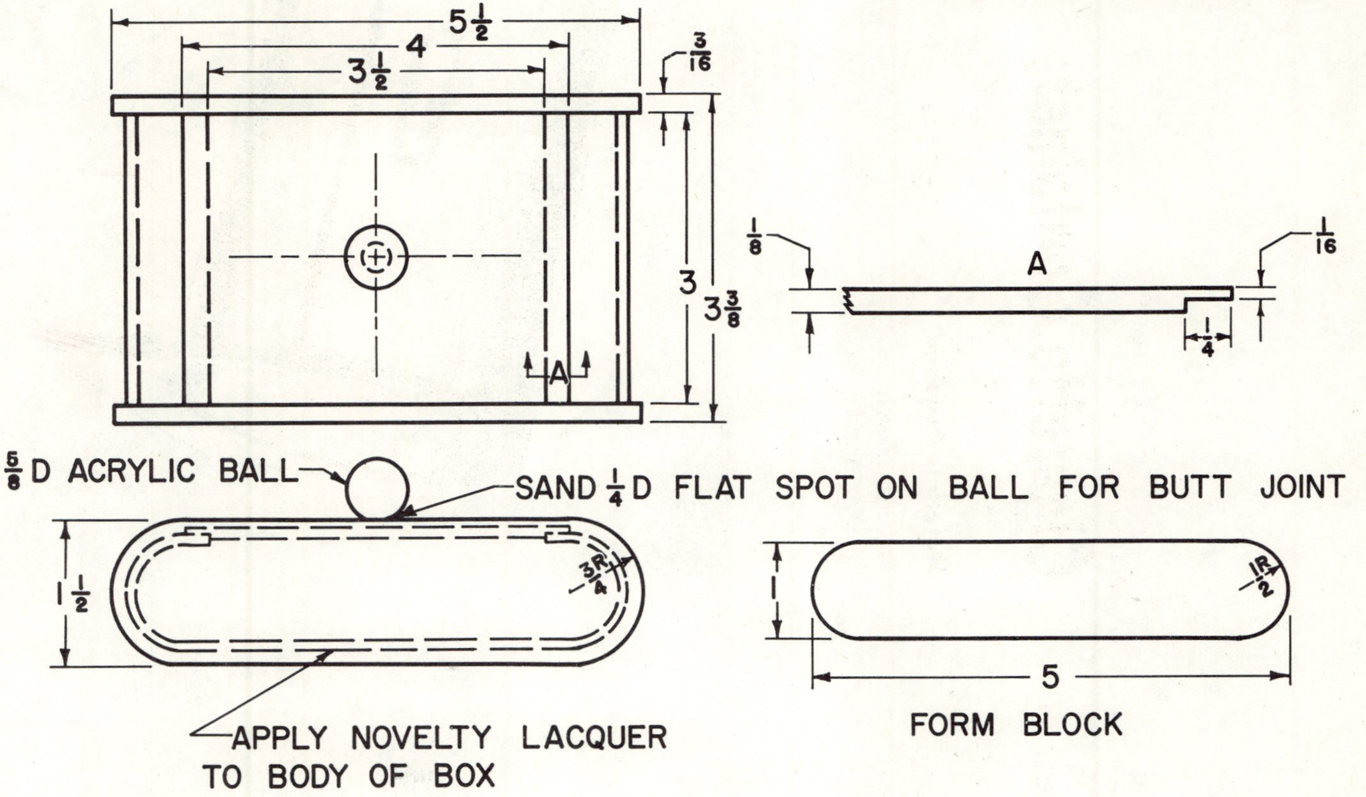

FORM BLOCK

PROJECT 19

Cigarette Container

Materials Required

1/8 Clear Acrylic
1/4 Dia. Acrylic Rod
Wire Brads

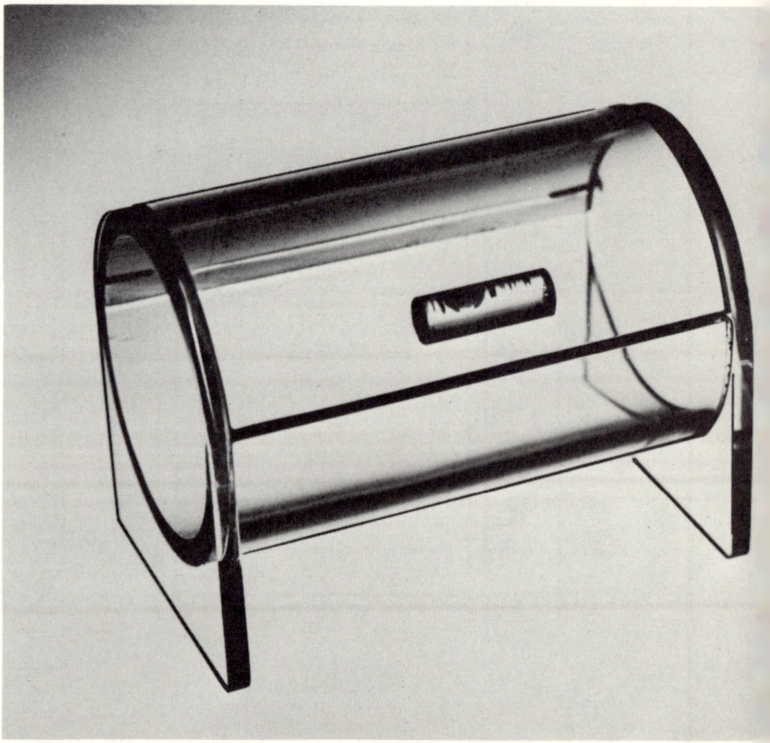

CIGARETTE CONTAINER

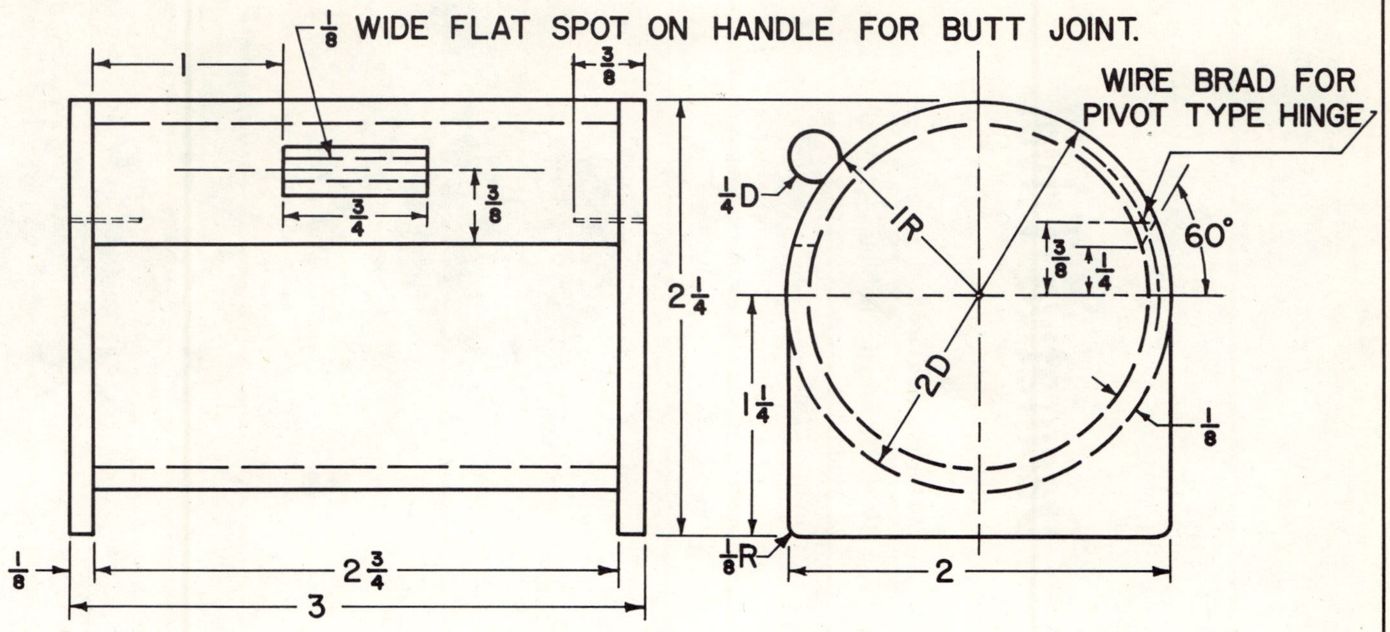

PROJECT 20

Decorative Trivet

Materials Required

3¼ O.D. Acrylic Tubing, 3/16 Wall Thickness
¼ Walnut Plywood
Bronze Paint

Refer to Technique 1 for information on sandblasting.

DECORATIVE TRIVET

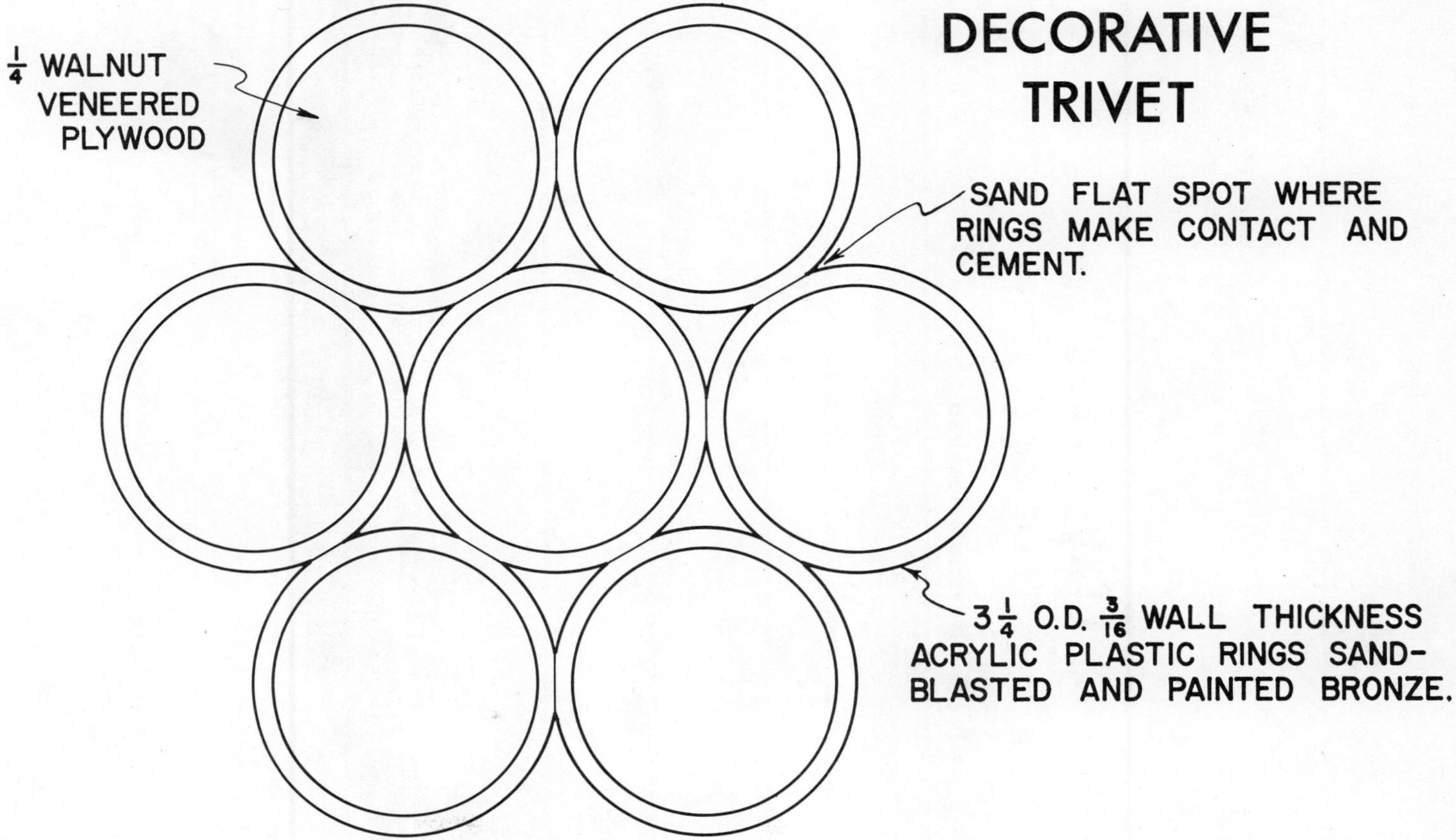

PROJECT 21

Tray

Materials Required

¼ Plywood
1⁄16 Plastic Laminate
⅜ Dia. Nylon Cord

By Allen Davis

SERVING TRAY

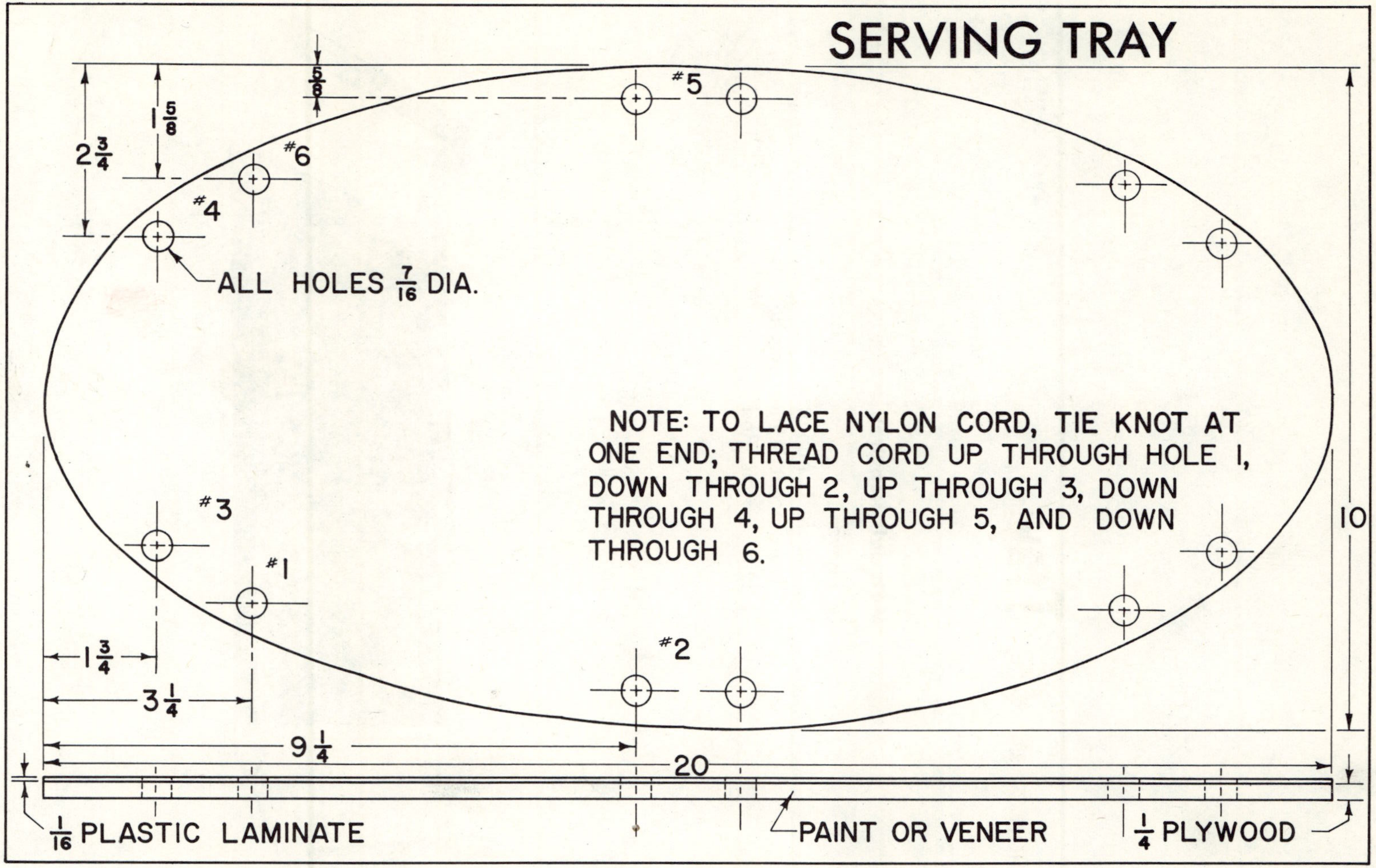

PROJECT 22

Tray

Materials Required

¼ Plywood
$\frac{1}{16}$ Plastic Laminate
¼ Translucent White Acrylic
½ Dia. Acrylic Rod

By Donald Shell

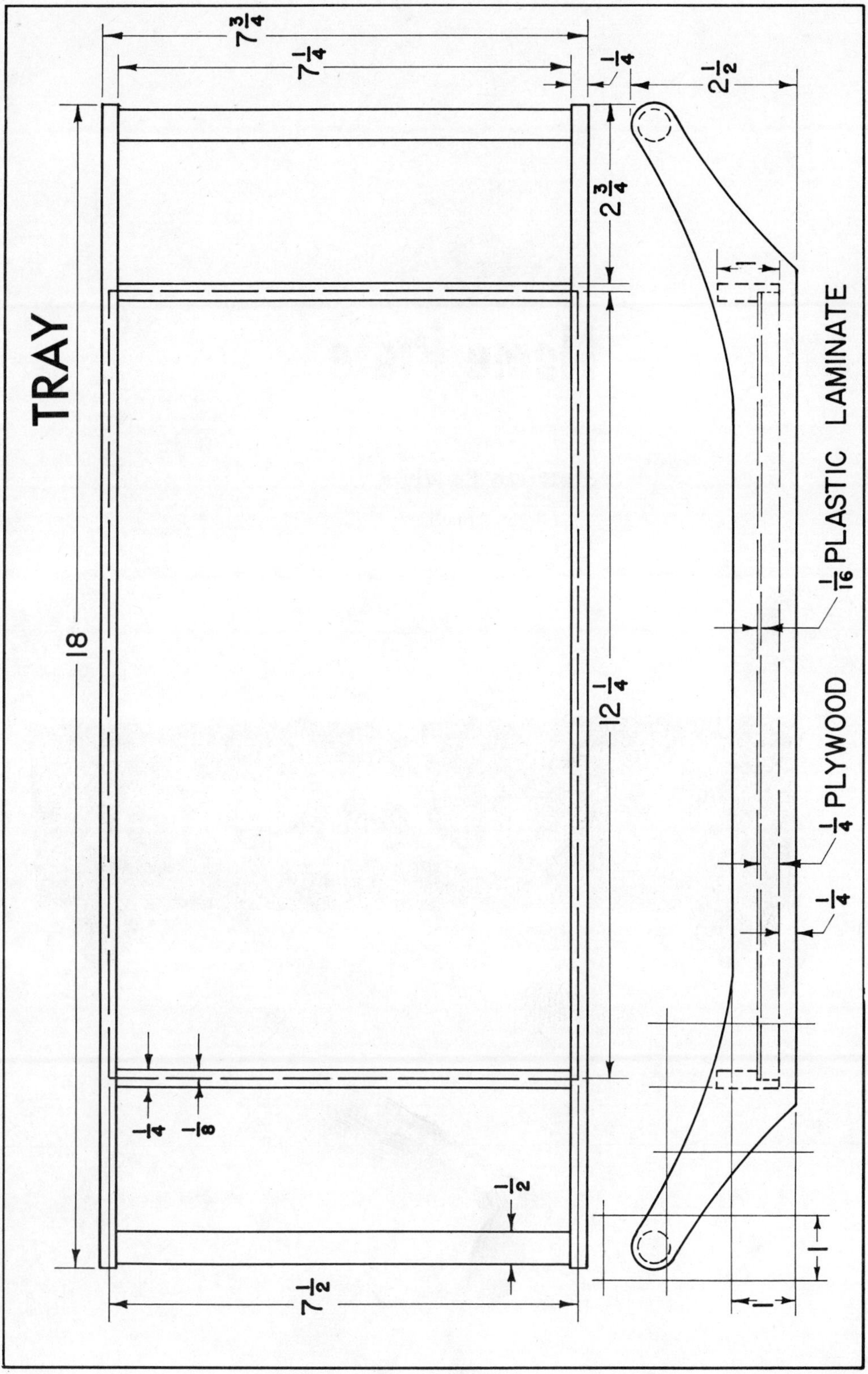
TRAY
18
7 3/4
7 1/4
1/4
2 1/2
2 3/4
12 1/4
1/4
1/8
1/2
7 1/2
1/16 PLASTIC LAMINATE
1/4 PLYWOOD
1/4
1
1

PROJECT 23

Name Plate

Materials Required

¼ Clear Acrylic
⅜ Opaque Green Acrylic

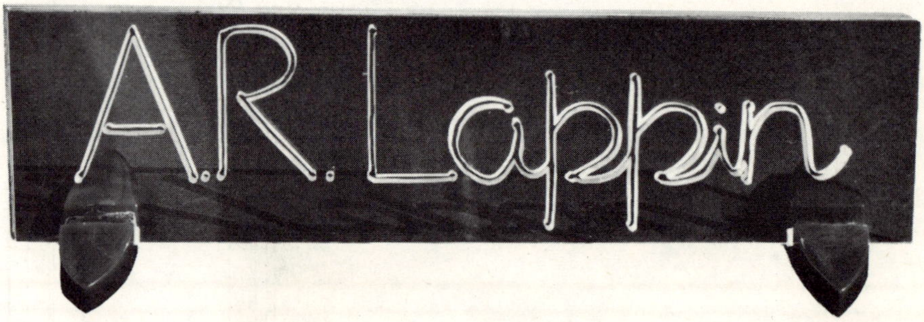

NAME PLATE

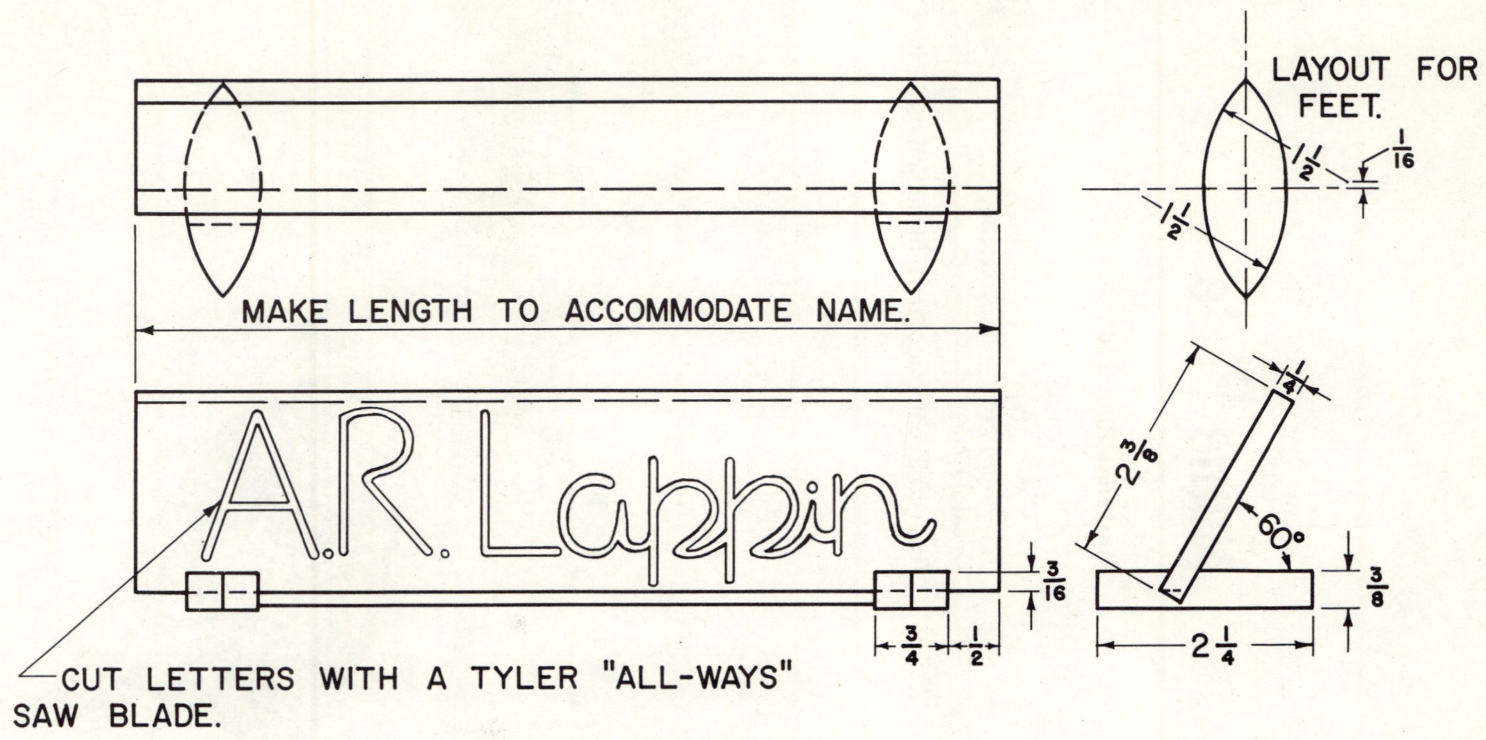

LAYOUT FOR FEET.
1 1/2
1 1/2
1/16
MAKE LENGTH TO ACCOMMODATE NAME.
A.R. Lappin
3/16
3/4
1/2
1/4
2 3/8
60°
3/8
2 1/4
CUT LETTERS WITH A TYLER "ALL-WAYS" SAW BLADE.

PROJECT 24

Name Plate

Materials Required

¼ Clear Acrylic
³⁄₁₆ Clear Acrylic
³⁄₁₆ Transparent Blue Acrylic
Decal Letters

Refer to Technique 2 for information on heat sealing.

NAME PLATE

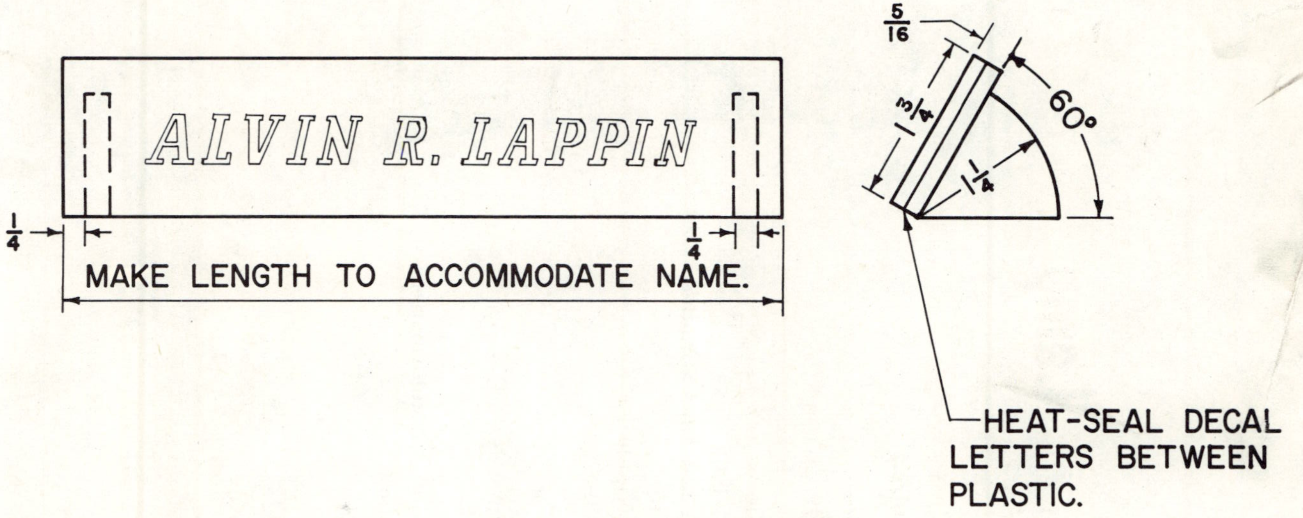

ALVIN R. LAPPIN
1/4
1/4
MAKE LENGTH TO ACCOMMODATE NAME.
5/16
3/4
1/4
60°
HEAT-SEAL DECAL LETTERS BETWEEN PLASTIC.

PROJECT 25

Lamp

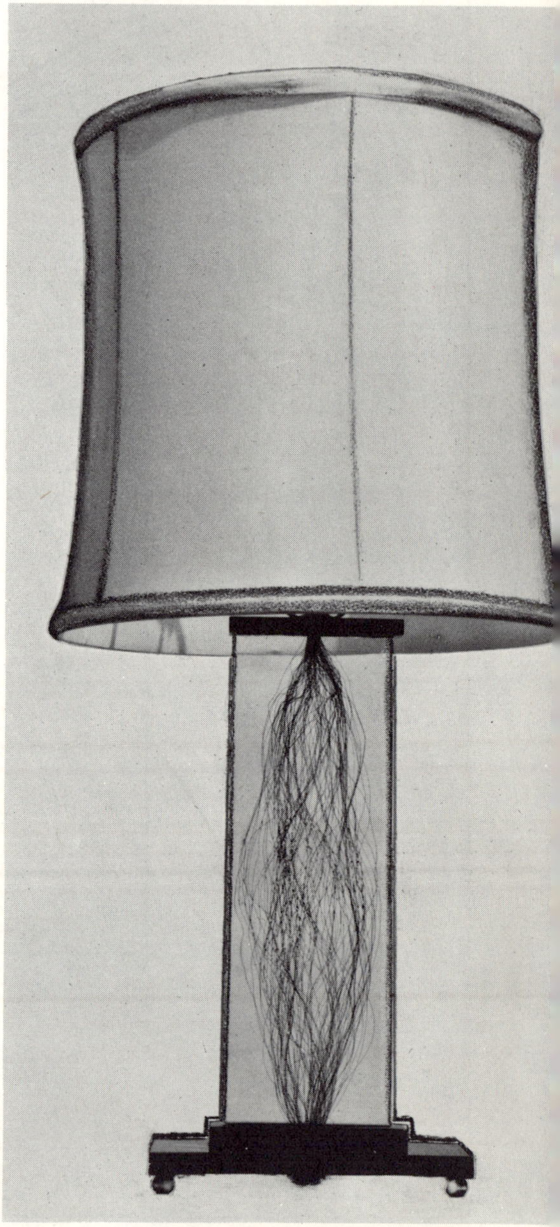

Materials Required

½ Black Acrylic
⅜ Black Acrylic
¼ Clear Acrylic
3/16 Clear Acrylic
⅜ Dia. Acrylic Balls
Electrical Cord, 18-Gauge Stranded Copper Wire
⅜ Dia. Pipe (Nipple)
Electrical Switch-Socket
Harp, 10-Inch
Finial
Shade (Approximately 12 Dia. by 12 High.)

Refer to Technique 2 for information on heat sealing.

By Robert Larson

LAMP

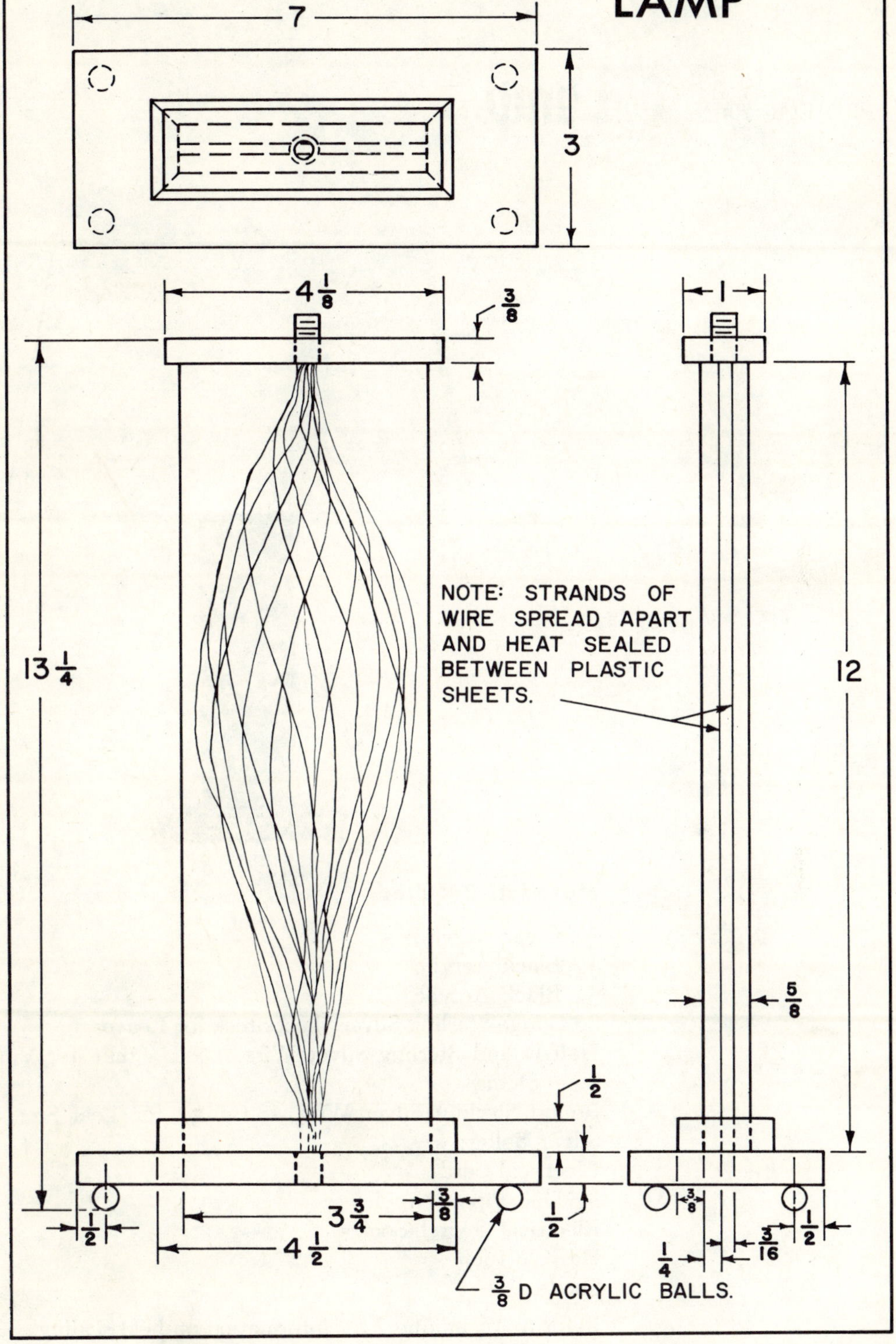
7
3
4 1/8
3/8
13 1/4
NOTE: STRANDS OF WIRE SPREAD APART AND HEAT SEALED BETWEEN PLASTIC SHEETS.
1
12
5/8
1/2
3/8
1/2
3 3/4
4 1/2
3/8 D ACRYLIC BALLS.
1/4
3/16

PROJECT 26

Lamp

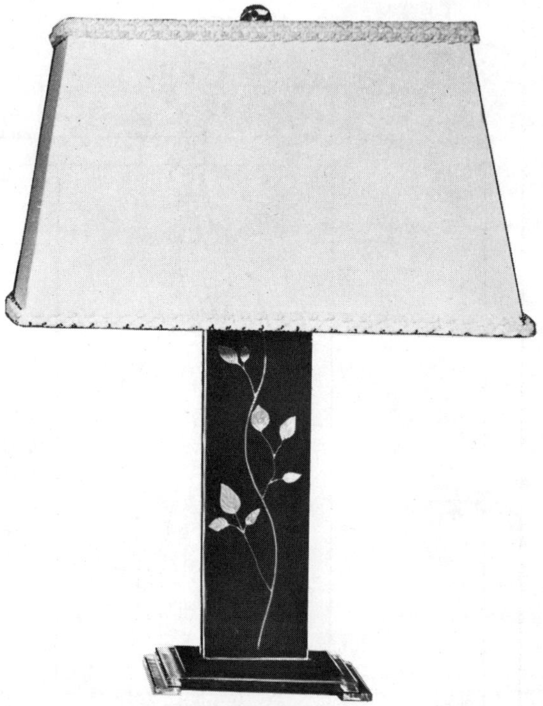

Materials Required

$\frac{3}{8}$ Clear Acrylic
$\frac{1}{4}$ Black Acrylic
$\frac{1}{8}$ Black Acrylic
36-Gauge Sterling Silver Sheet Stock for Leaves
Half-Round Sterling Silver Wire, 4 x $1\frac{1}{2}$ mm, for Vine Stems
Round Sterling Silver Wire, 18-Gauge, for Leaf Stems
Silver Solder
Electrical Cord
$\frac{3}{8}$ Dia. Pipe (Nipple)
Electrical Switch-Socket
Harp, 8-Inch
Shade

Refer to Technique 2 for information on heat sealing.

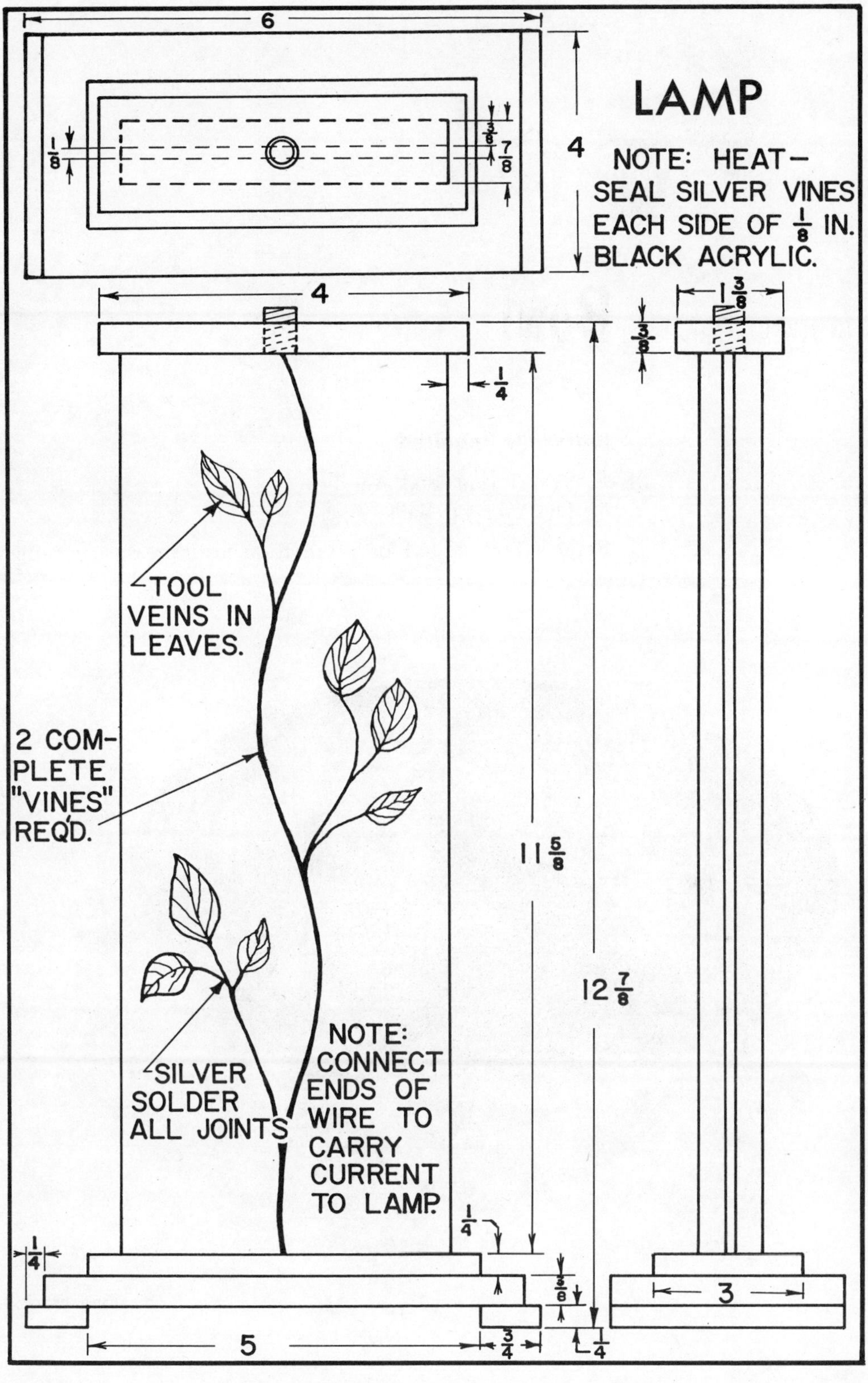

LAMP
NOTE: HEAT-SEAL SILVER VINES EACH SIDE OF 1/8 IN. BLACK ACRYLIC.
6
4
1/8
3/8
7/8
4
1 3/8
3/8
1/4
TOOL VEINS IN LEAVES.
2 COMPLETE "VINES" REQ'D.
11 5/8
12 7/8
SILVER SOLDER ALL JOINTS
NOTE: CONNECT ENDS OF WIRE TO CARRY CURRENT TO LAMP.
1/4
1/4
3/8
3
5
3/4
1/4

PROJECT 27

Bowl

Materials Required

1/8 Translucent Red Acrylic
5/8 Dia. Acrylic Balls

Refer to Technique 4 for information on air pressure forming.

By John Brugo

BOWL

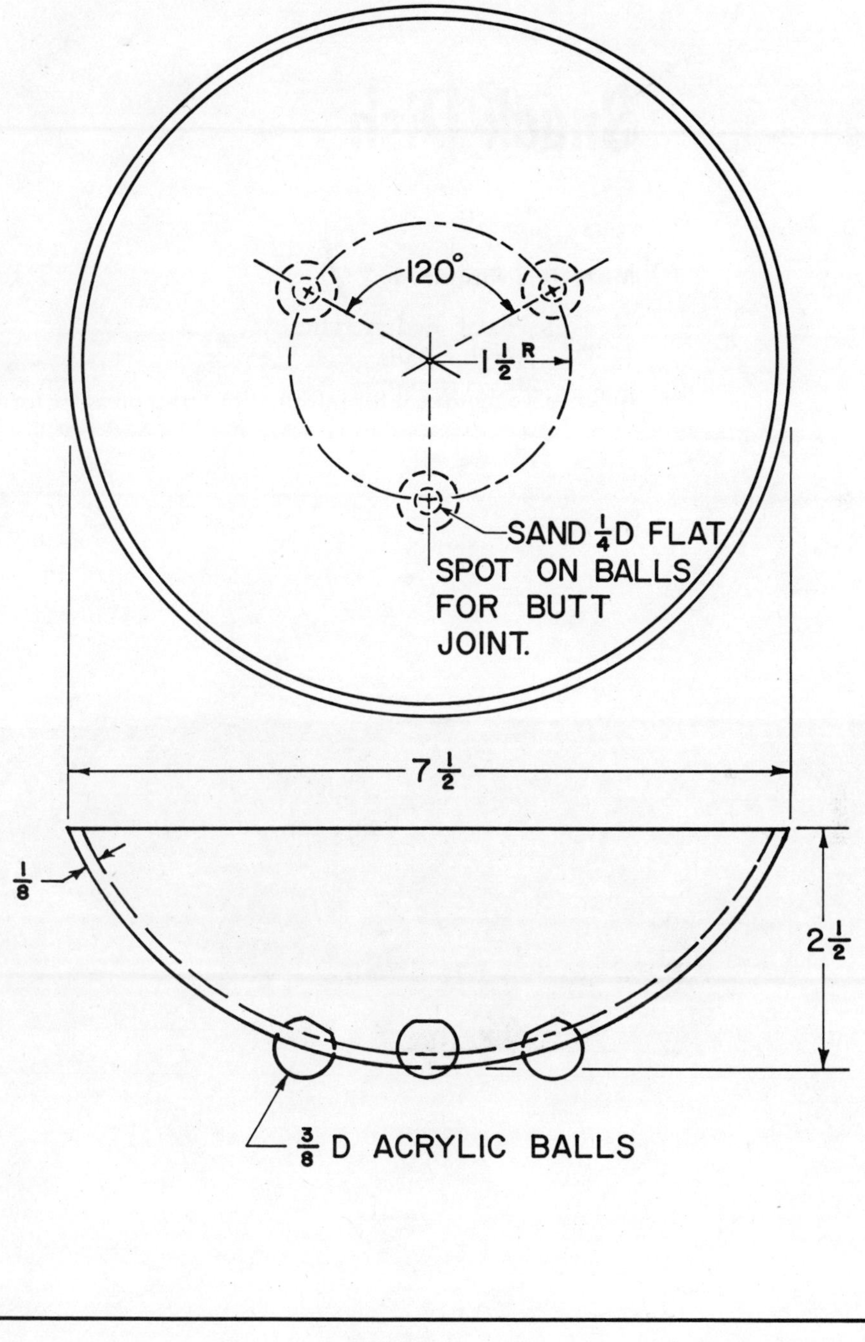
120°
1 1/2 R
SAND 1/4 D FLAT
SPOT ON BALLS
FOR BUTT
JOINT.
7 1/2
1/8
2 1/2
3/8 D ACRYLIC BALLS

PROJECT 28

Snack Dish

Materials Required

1/8 Translucent Red Acrylic
1/2 Dia. Acrylic Balls

Refer to Technique 4 for information on air pressure forming.

By Robert Larson

SNACK DISH

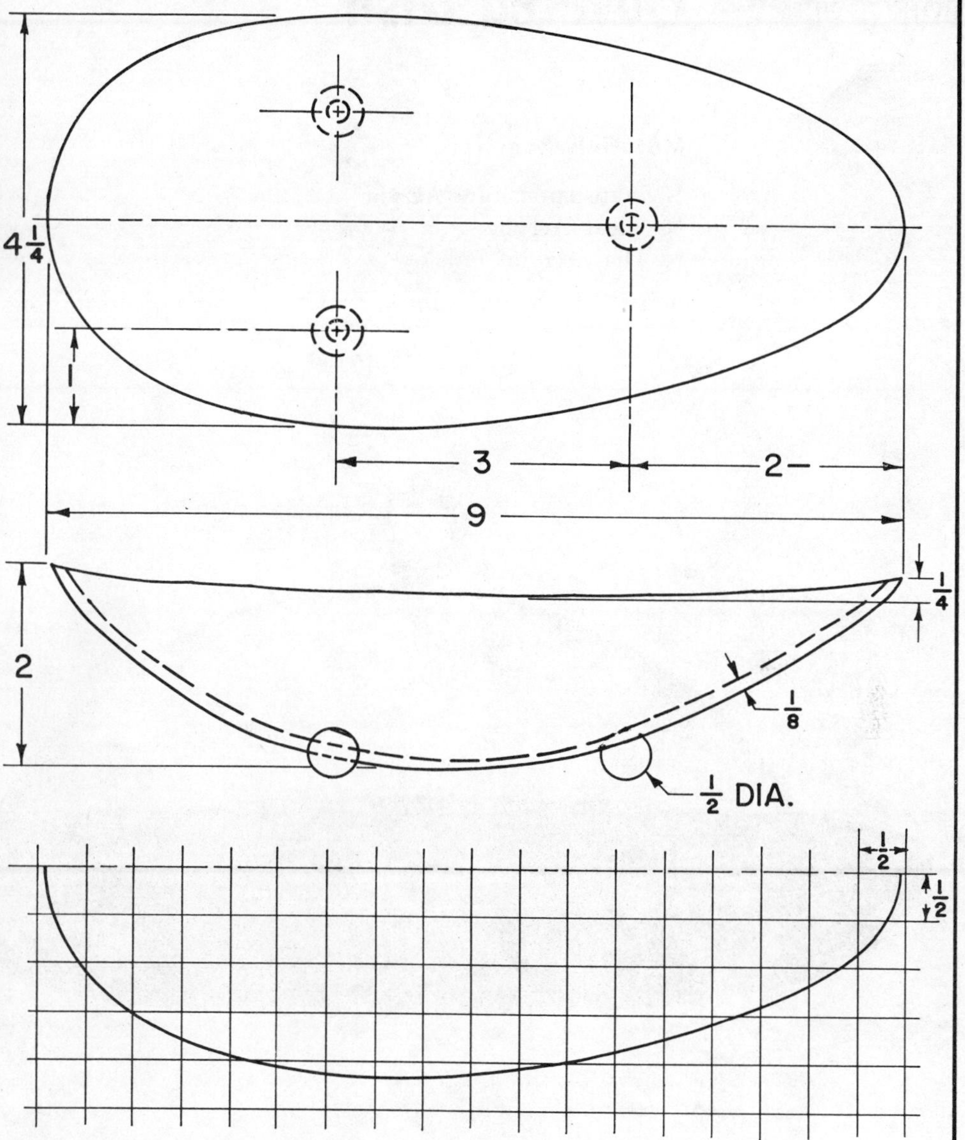

LAYOUT FOR BLOW FORMING JIG.

PROJECT 29

Covered Dish

Materials Required

1/8 Transparent Blue Acrylic
1/8 Clear Acrylic
5/8 Dia. Acrylic Ball

Refer to Technique 4 for information on air pressure forming.

COVERED DISH

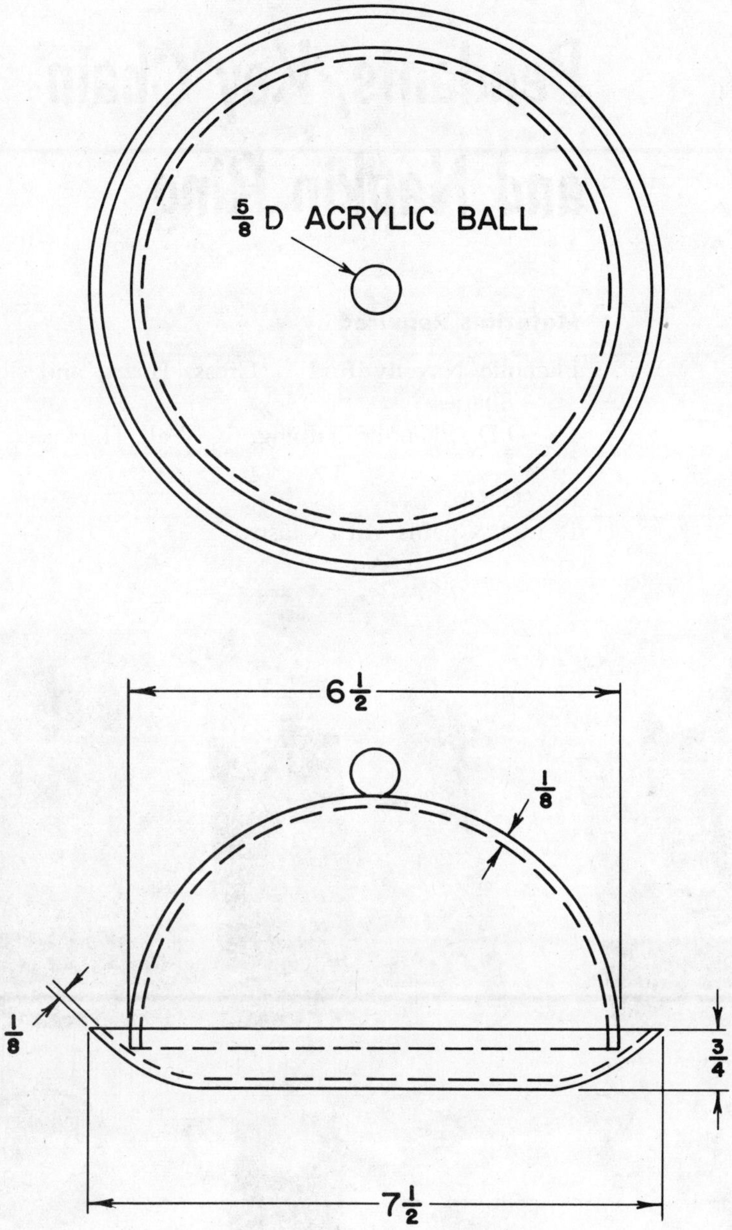

PROJECT 30

Pendants, Key Chain and Napkin Ring

Materials Required

Phenolic Novelty Rod — Cross, Heart, and Scotty Dog Shapes
1½ O.D. Phenolic Tubing, ⅛ Wall Thickness
Eyes
Key Chains
18-Inch Chains with Clasp

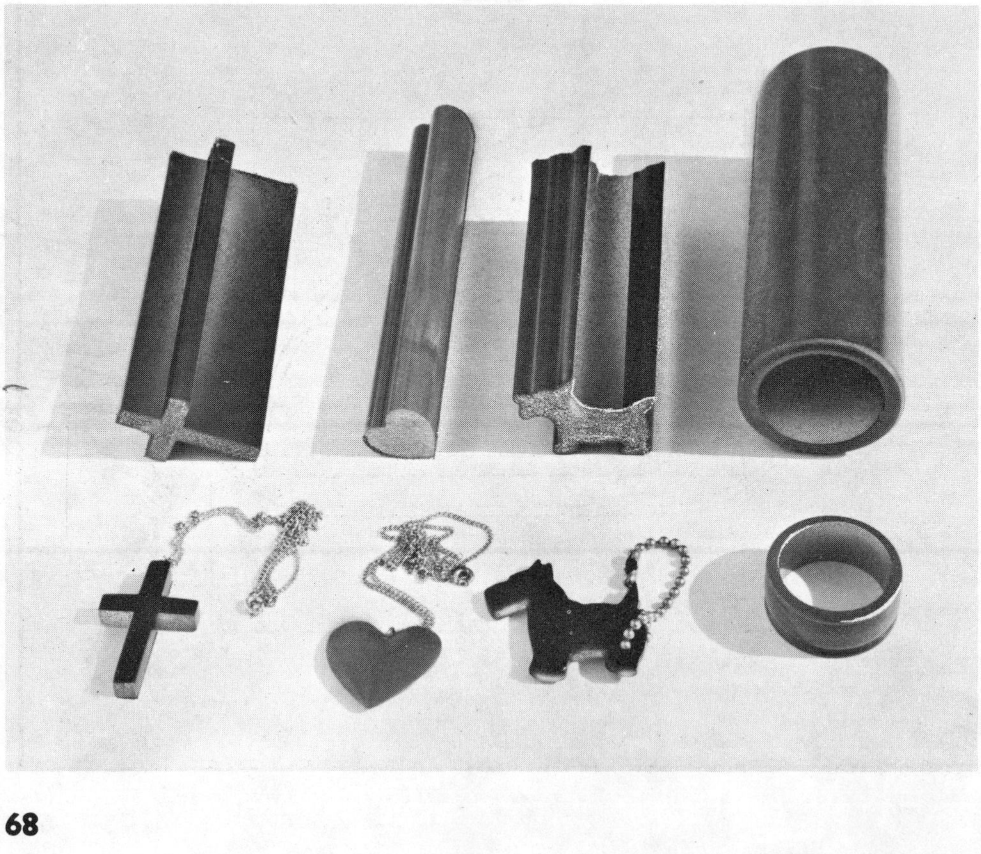

PENDANTS, KEY CHAIN AND NAPKIN RING

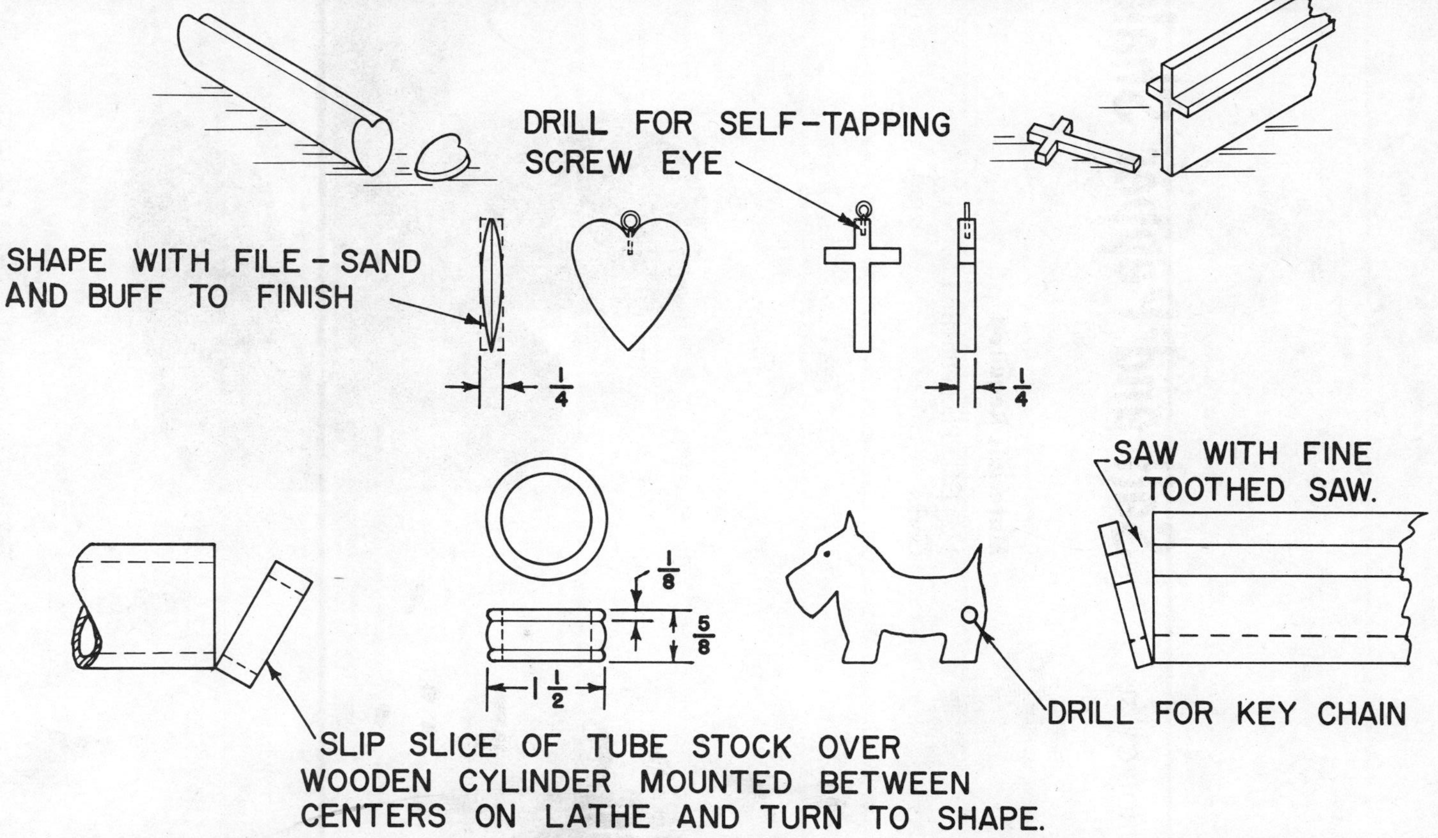

PROJECT 31

Salt and Pepper Shakers

Materials Required

$1\frac{1}{8}$ Dia. Phenolic Rod
Corks

This project appeared earlier in **School Shop.**

SALT AND PEPPER SHAKERS

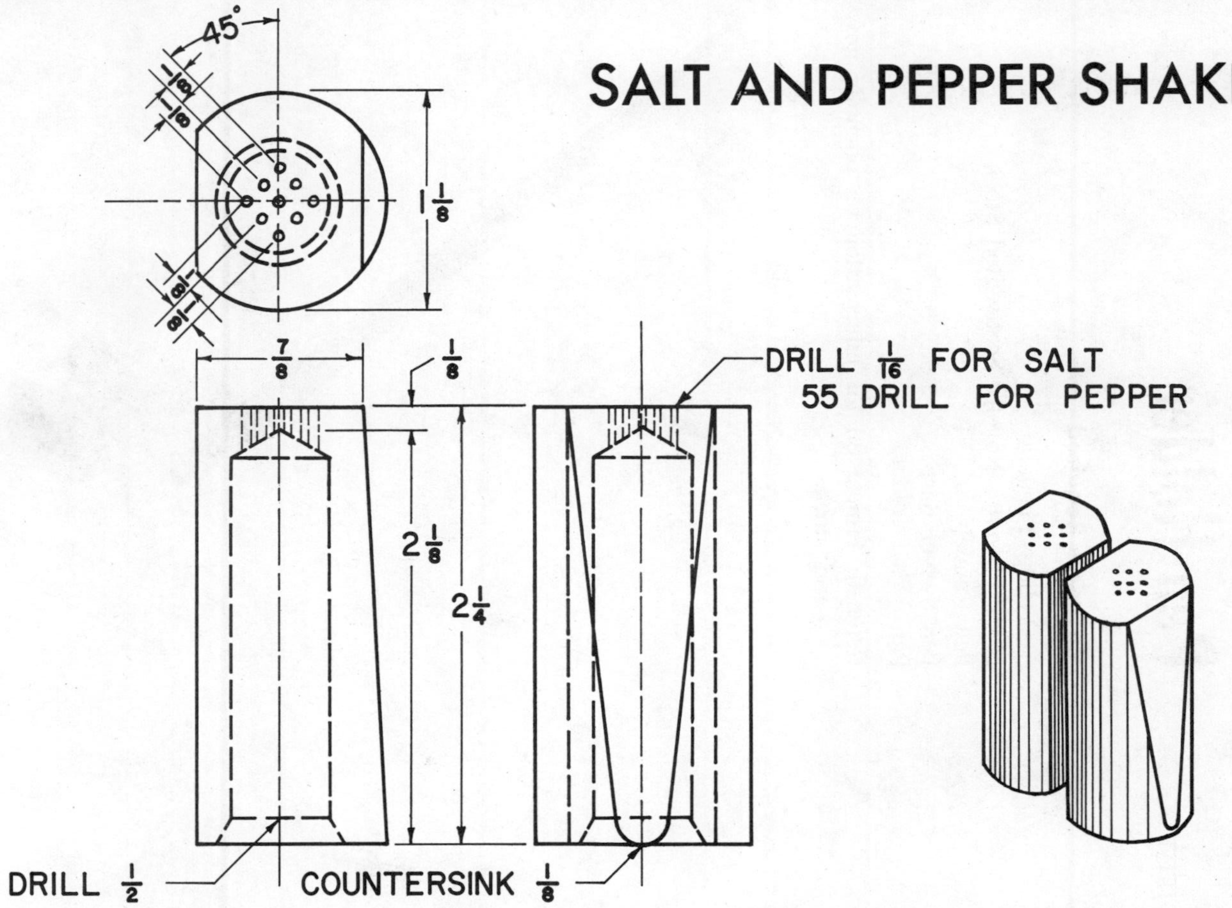

PROJECT 32

Pen Holder

Materials Required

3/8 Yellow Phenolic
Black Phenolic Novelty Scotty Rod Stock
Pen Funnel
Pen Swivel Base
Pen

Refer to Project 30 for information on phenolic novelty rod stock.

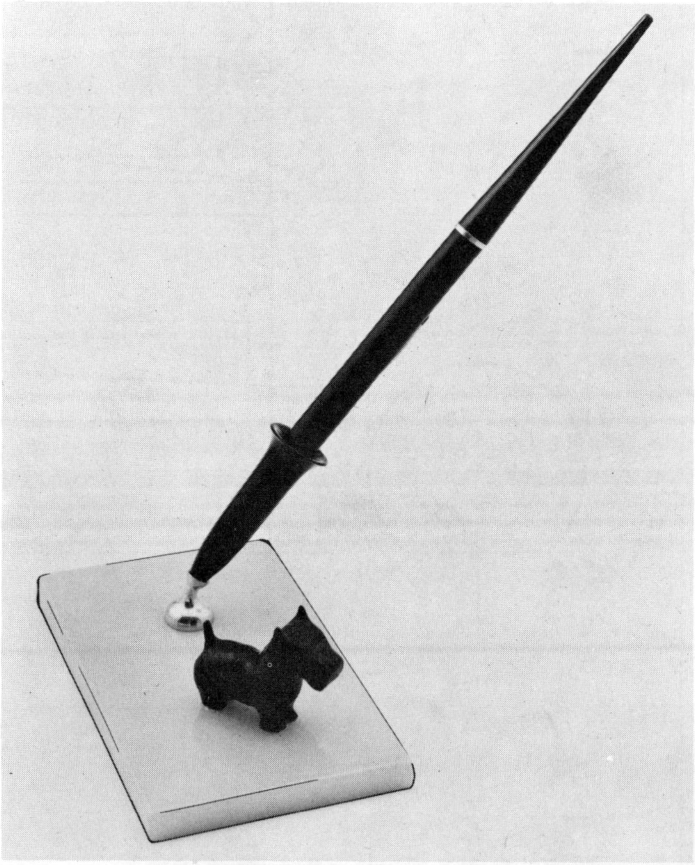

PEN HOLDER

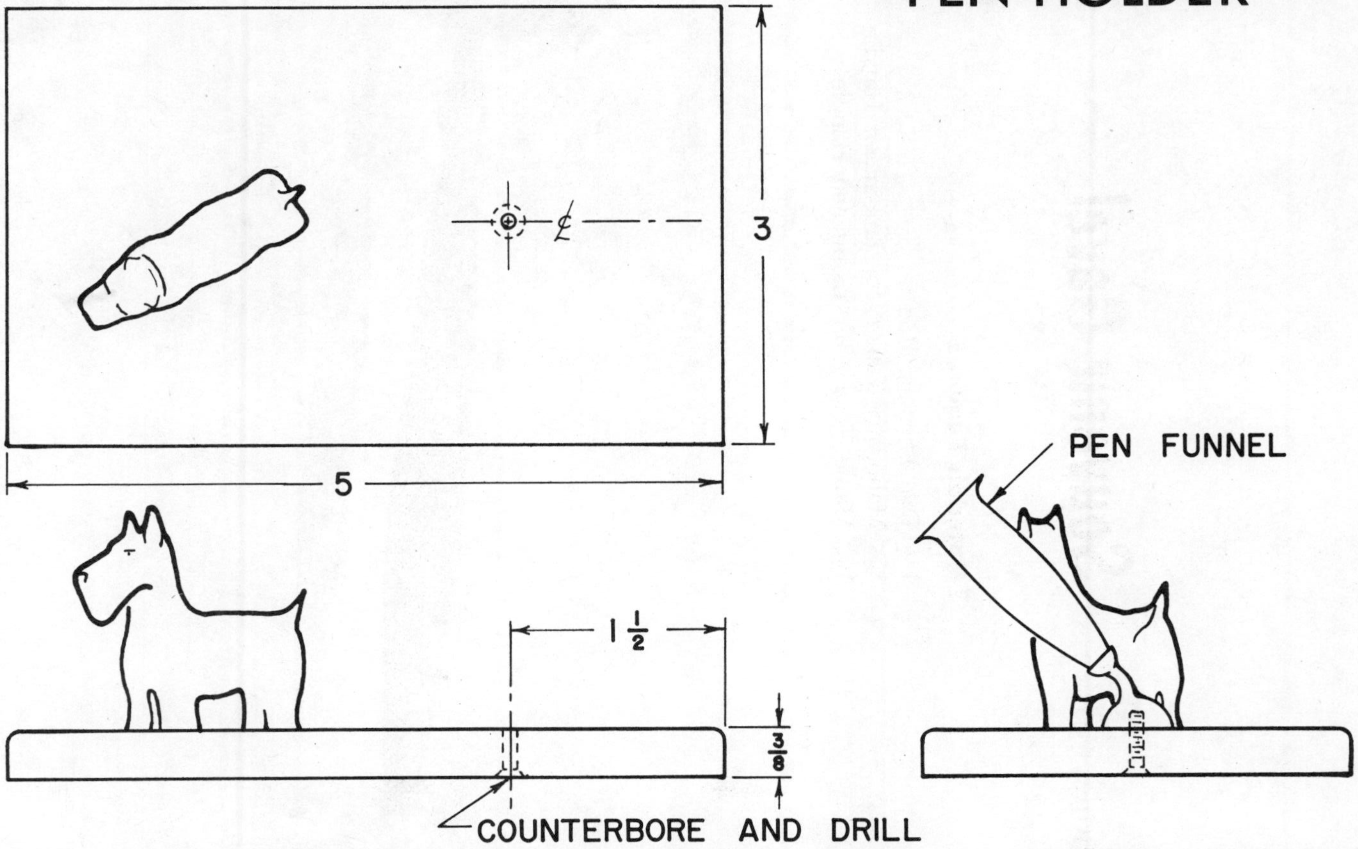

PROJECT 33

Souvenir Gavel

Materials Required

1½ Dia. Phenolic Rod
1/28 Mahogany Veneer for Laminated Handle
1/28 Maple Veneer for Laminated Handle

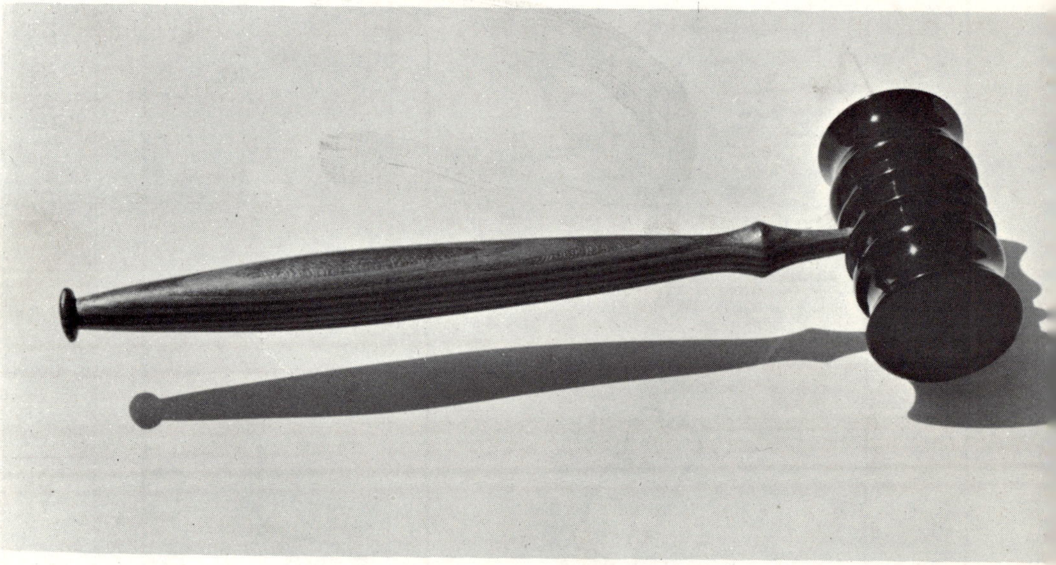

SOUVENIR GAVEL

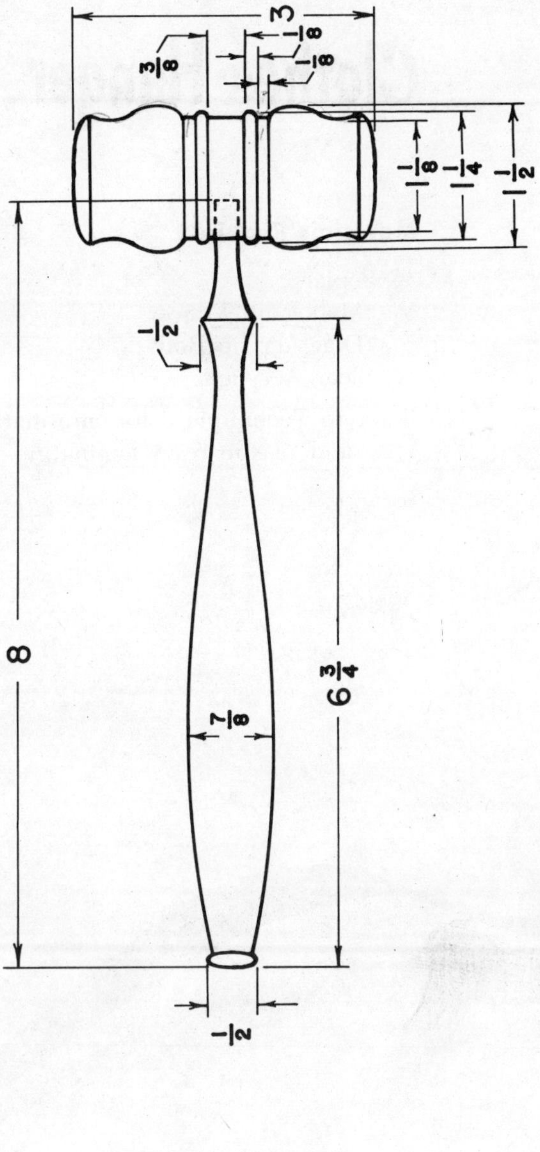

PROJECT 34

Clothes Hanger

Materials Required

$\frac{1}{8}$ Poplar
$\frac{1}{28}$ Mahogany Veneer
$\frac{5}{16}$ Dia. Acrylic Rod
$\frac{1}{8}$ Clear Acrylic

Refer to Technique 6 for information on forms for wood and plastic resin laminating.

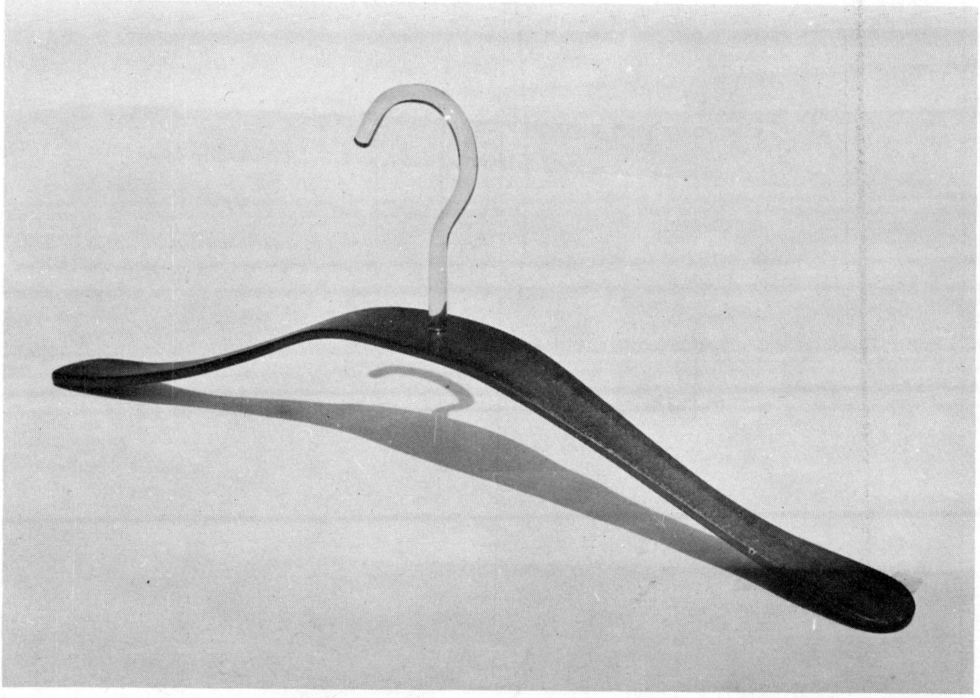

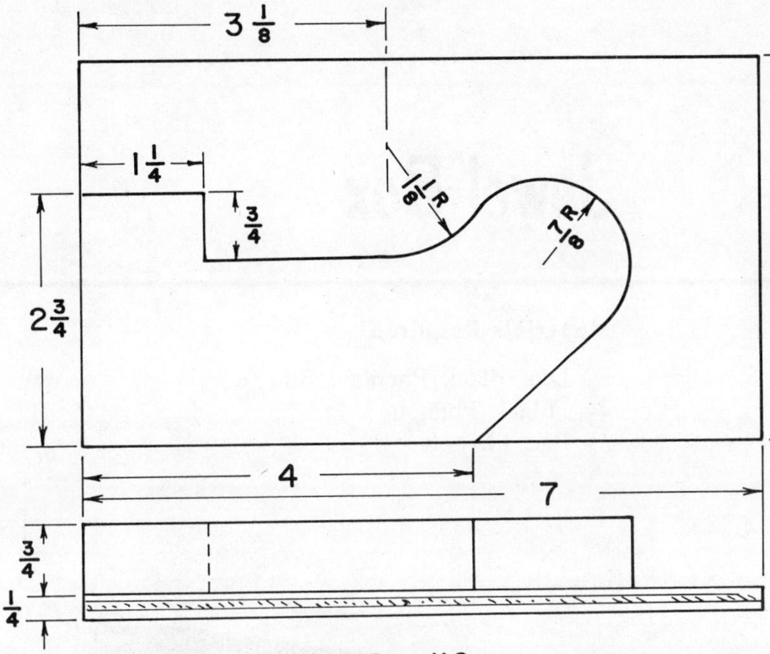

HOOK FORMING - JIG

CLOTHES HANGER

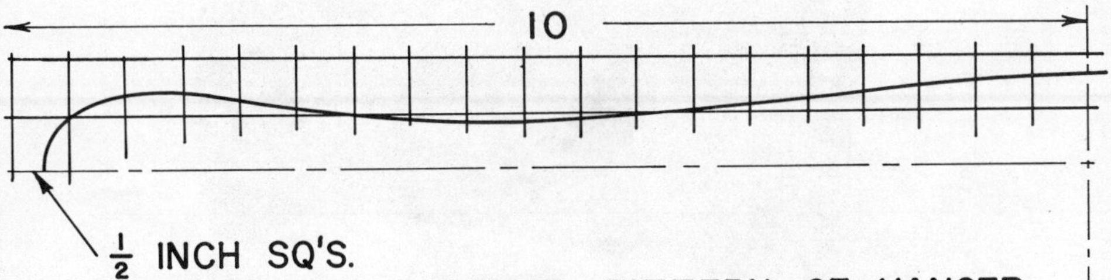

QUARTER PATTERN OF HANGER

PROJECT 35

Jewel Box

Materials Required

$\frac{5}{8}$ Dia. Black Phenolic Rod
$\frac{5}{16}$ Black Phenolic
$\frac{1}{8}$ Red Phenolic

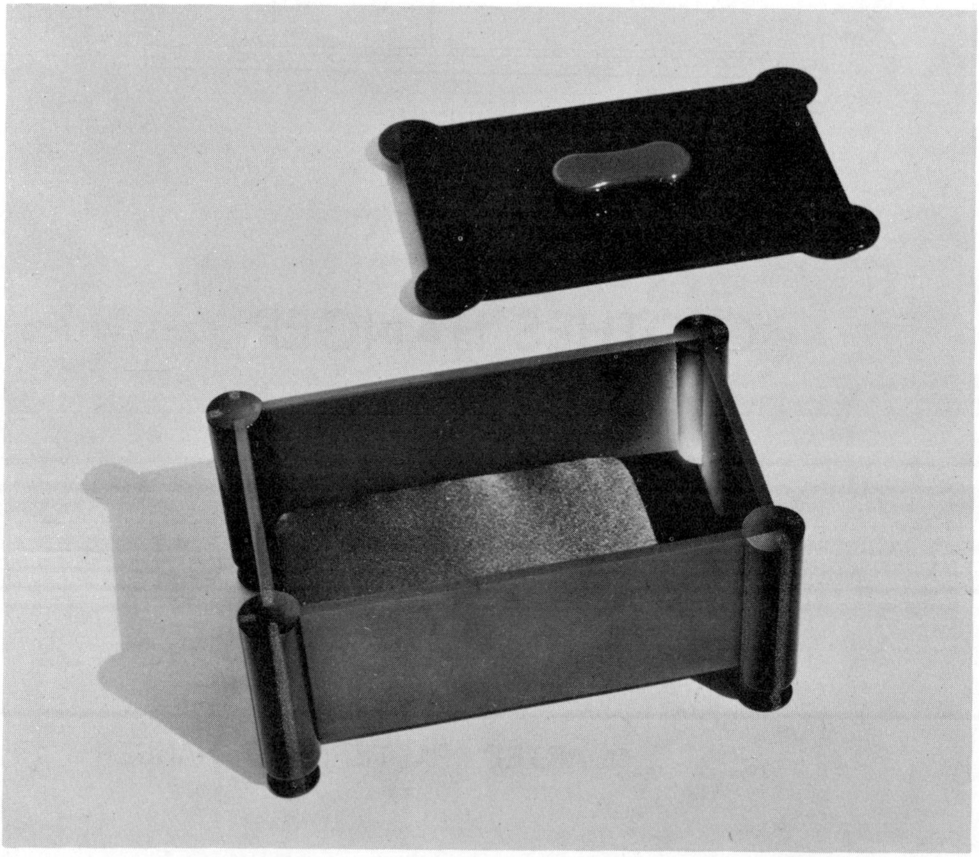

JEWEL BOX

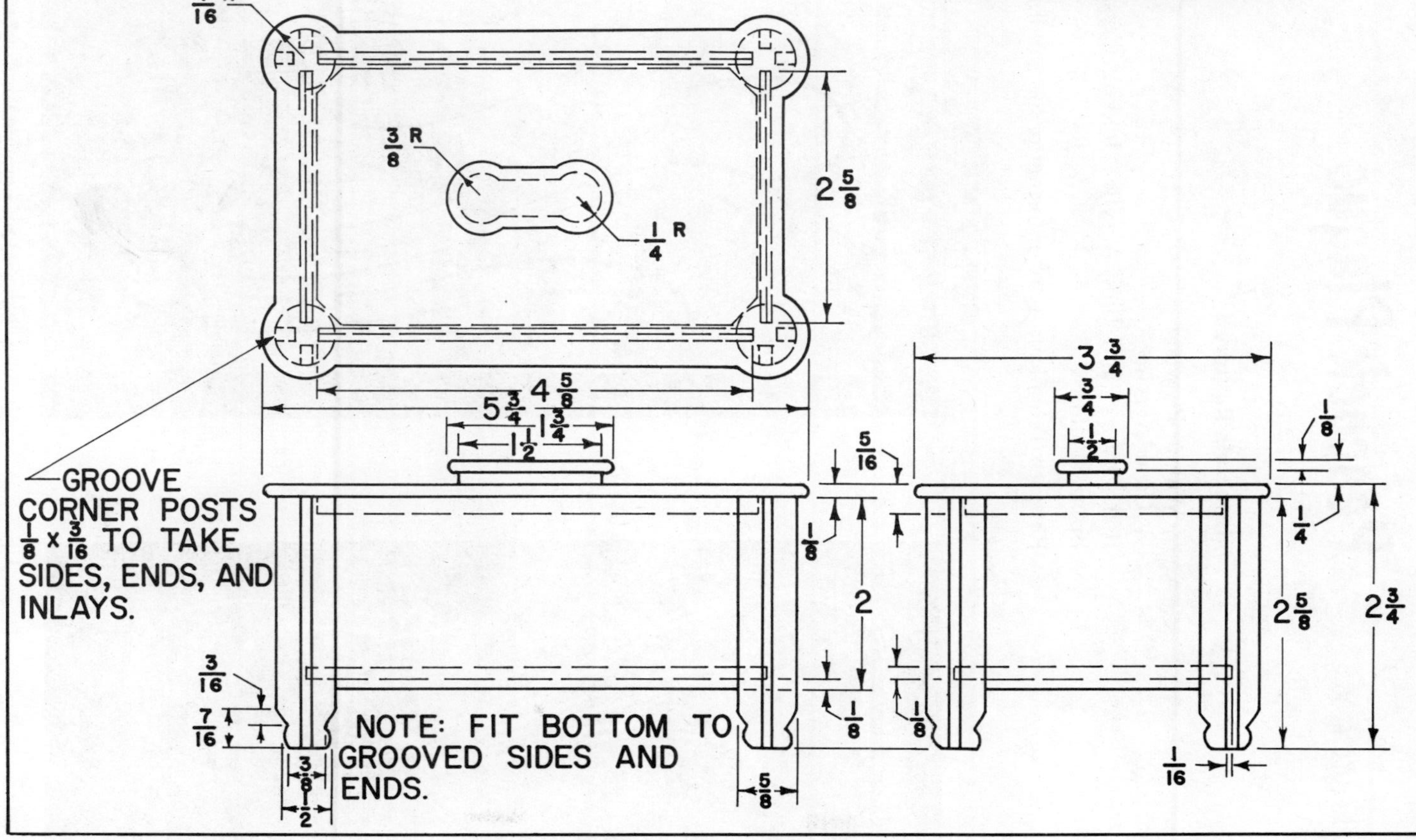

GROOVE CORNER POSTS 1/8 x 3/16 TO TAKE SIDES, ENDS, AND INLAYS.
NOTE: FIT BOTTOM TO GROOVED SIDES AND ENDS.
7/16 R
3/8 R
1/4 R
2 5/8
5 3/4
4 5/8
1 3/4
1 1/2
3 3/4
3/4
1/2
5/16
1/8
1/4
2
2 5/8
2 3/4
1/16
3/16
7/16
3/8
5/8

PROJECT 36

Peacock Plaque

Materials Required

¼ Walnut Plywood
¼ Clear Acrylic
Red, Orange, and Blue Acrylic Molding Crystals
Bronze Paint

Refer to Technique 1 for information on sandblasting.

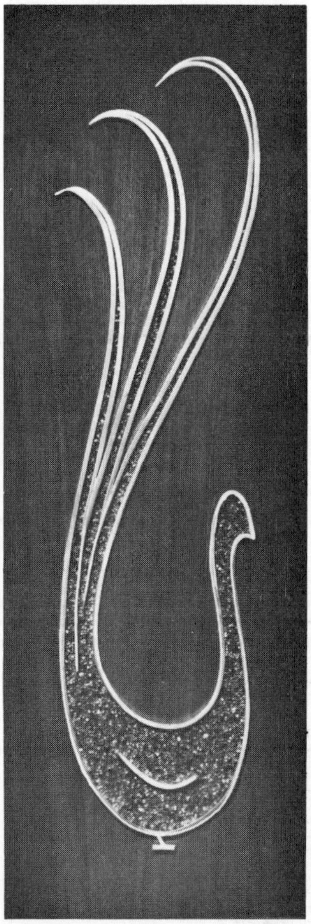

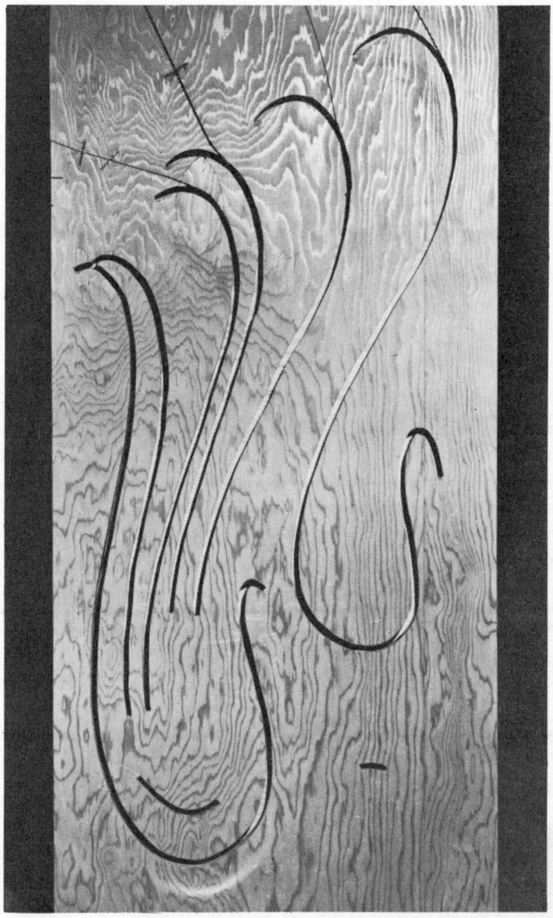

Jig (¼″ Plywood) for Forming Parts of Peacock Plaque

PEACOCK PLAQUE

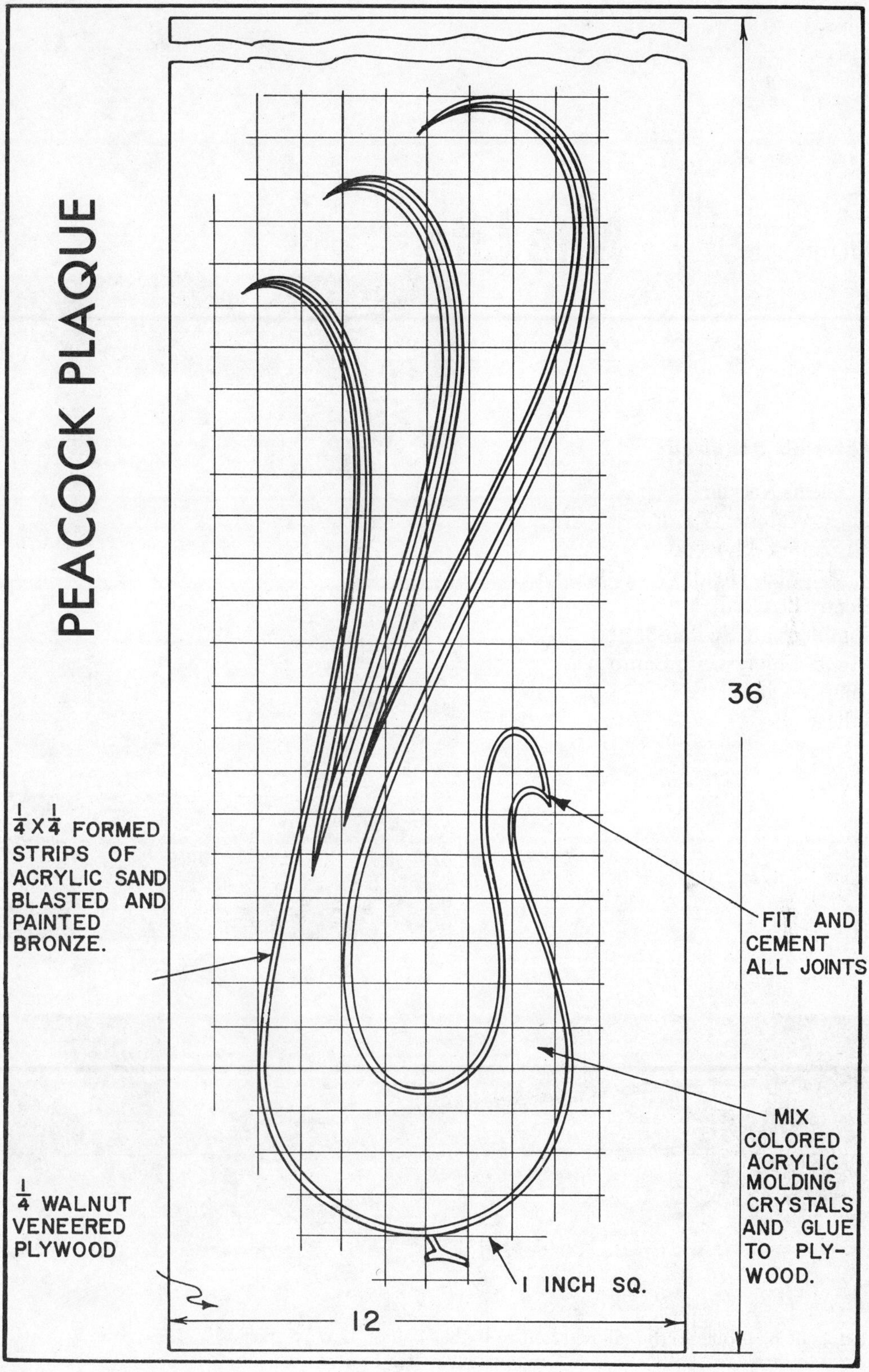

PROJECT 37

Nite Lite

Materials Required

¼ Clear Acrylic
Black Walnut Wood
¼ Walnut Plywood
¾ #6 Flat Head Wood Screws
Electrical Cord
Combination Switch-Starter
9-Inch Fluorescent Lamp
Lamp Sockets
Ballast

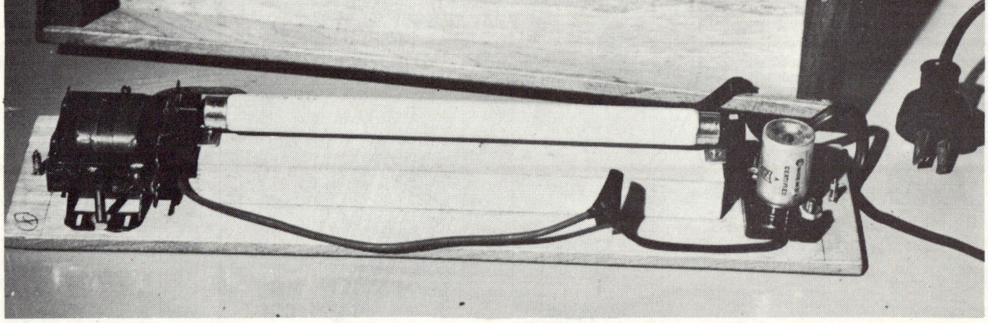

Bottom of Base Removed to Show (Left to Right) Ballast, Lamp Sockets with Lamp Installed, and Combination Switch-Starter

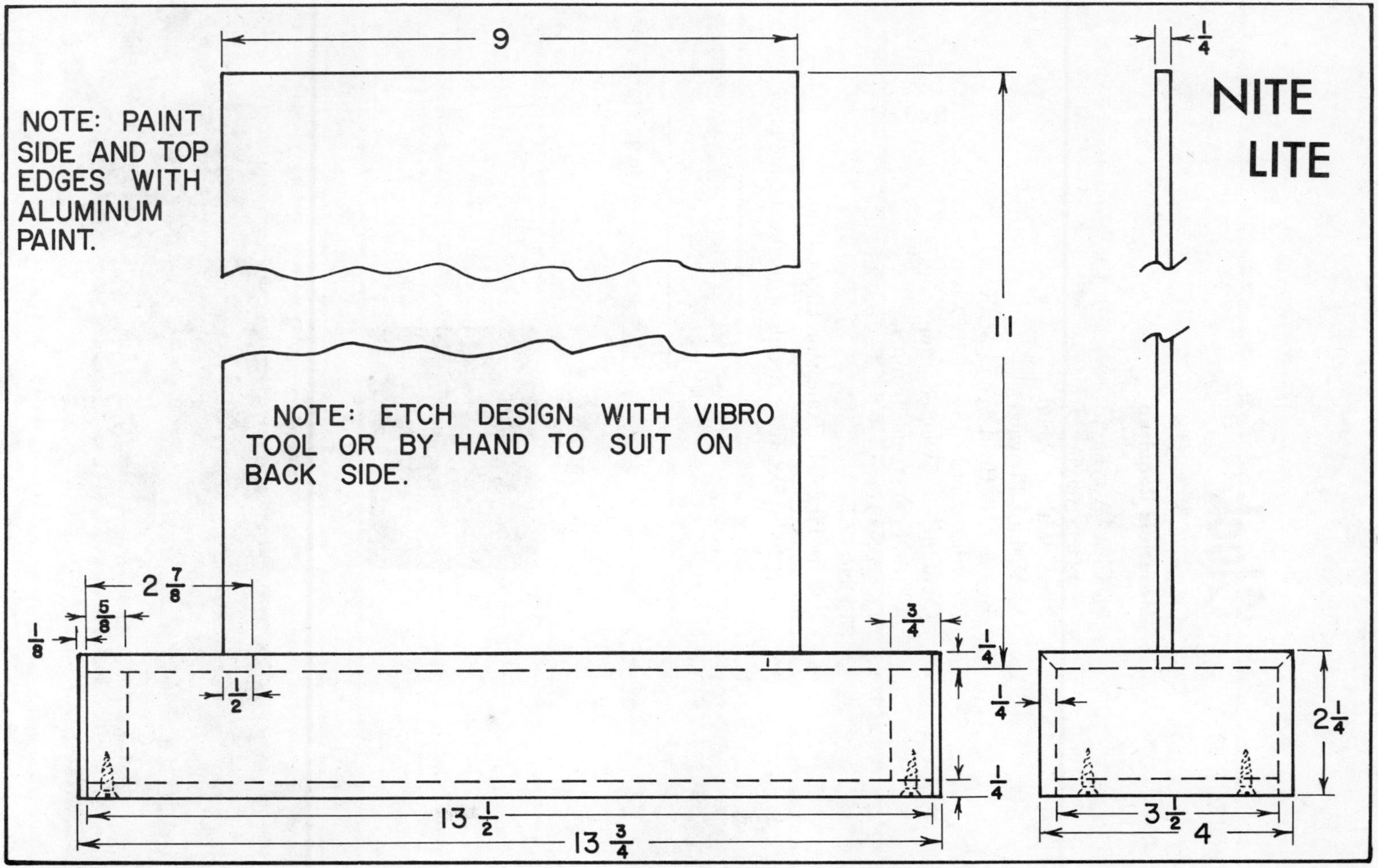
NITE
LITE
NOTE: PAINT SIDE AND TOP EDGES WITH ALUMINUM PAINT.
NOTE: ETCH DESIGN WITH VIBRO TOOL OR BY HAND TO SUIT ON BACK SIDE.
9
11
1/4
2 7/8
5/8
1/8
1/2
3/4
13 1/2
13 3/4
2 1/4
3 1/2
4

PROJECT 38

Clock

Materials Required

¼ Clear Acrylic
⅛ Black Acrylic
Black Walnut Wood
¼ Walnut Plywood
¾ #6 Flat Head Wood Screws
Electrical Cord
Combination Switch-Starter
9-Inch Fluorescent Lamp
Lamp Sockets
Ballast
Battery-Powered Transistor Clock Unit

Refer to Project 37 for details of enclosed lamp base.

By A. Charles Tyler

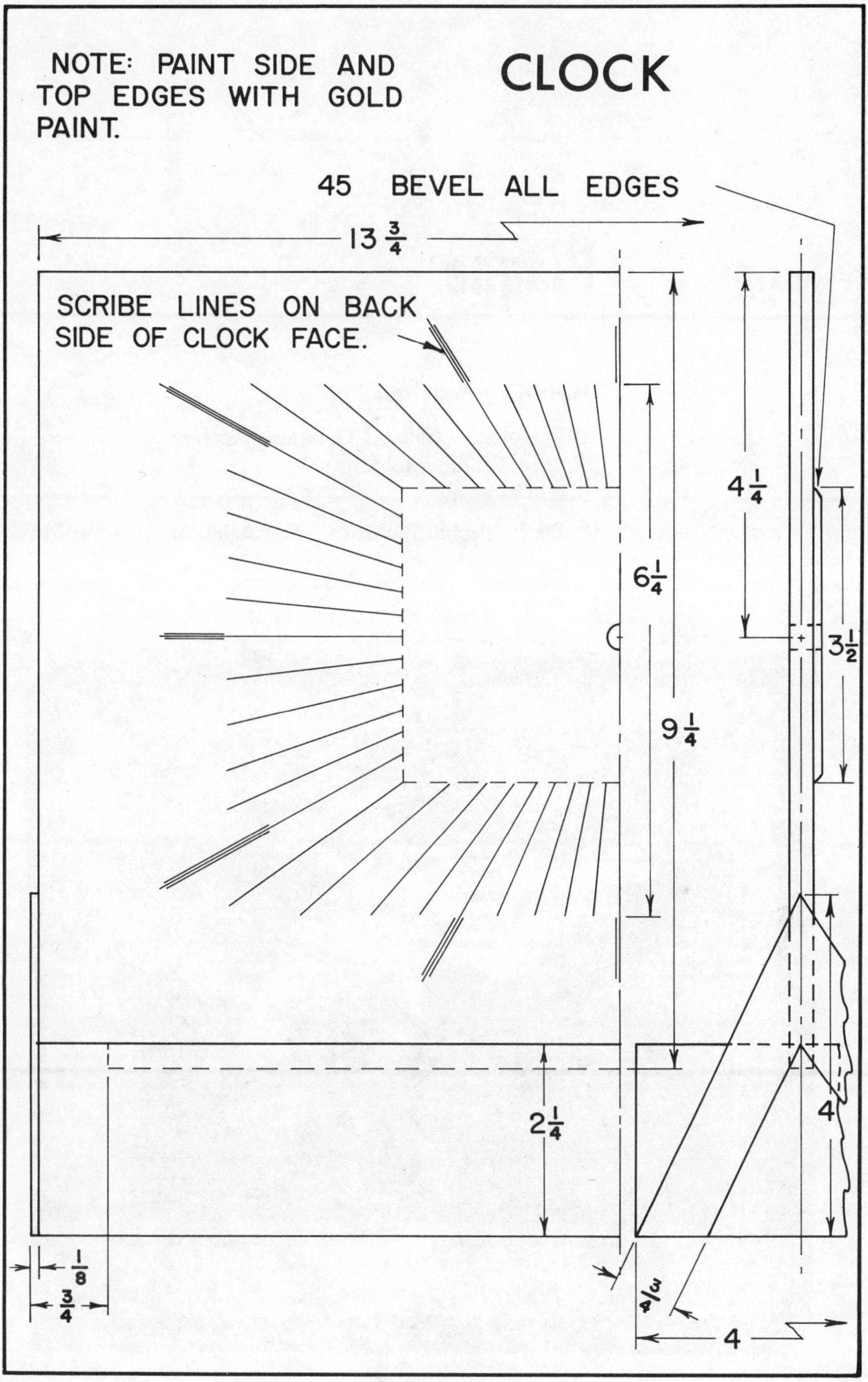

NOTE: PAINT SIDE AND TOP EDGES WITH GOLD PAINT.
CLOCK
45 BEVEL ALL EDGES
13 3/4
SCRIBE LINES ON BACK SIDE OF CLOCK FACE.
4 1/4
6 1/4
3 1/2
9 1/4
2 1/4
4
1/8
3/4
3/4
4

PROJECT 39

Plaque

Materials Required

.010 Gold Metallized Cellulose Acetate
Picture Frame Molding
Black Flocking

Refer to Technique 5 for information on vacuum forming.

PLAQUE

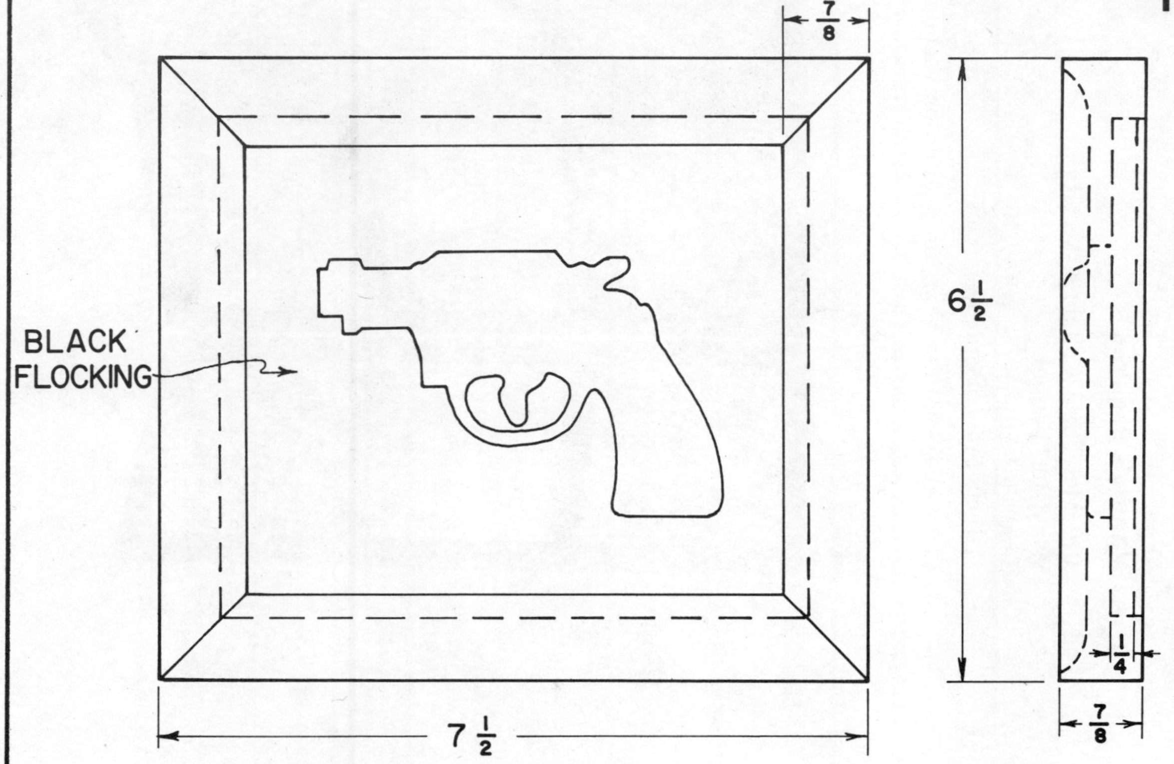

PROJECT 40

Decorative Screen

Materials Required

.010 Metallized Cellulose Acetate
Wood for Frames
Suede for Hinges
¼-Inch Brass Escutcheon Pins

Refer to Technique 5 for information on vacuum forming.

DECORATIVE SCREEN

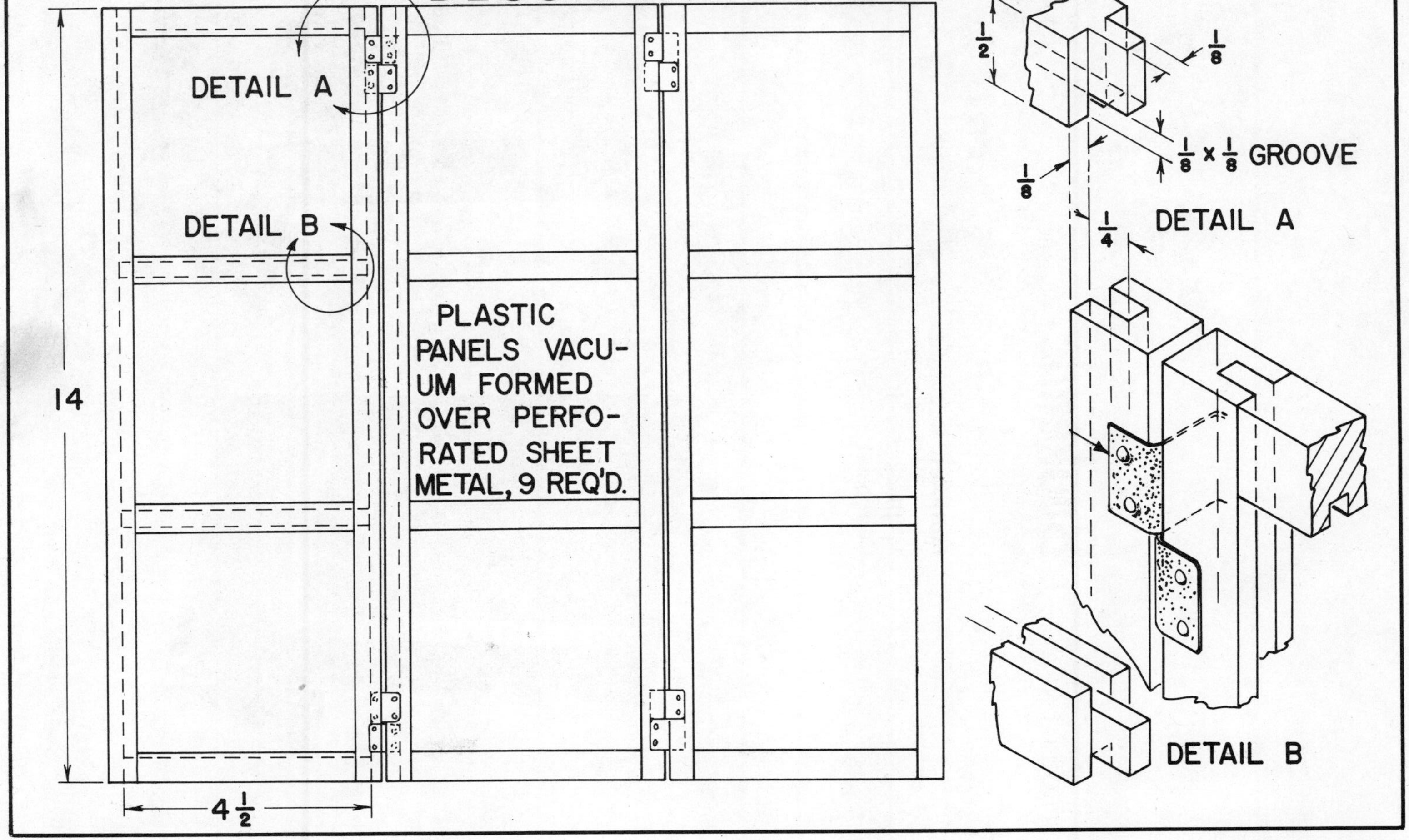

PROJECT 41

Bookends

Materials Required

¼ Black Acrylic
Cerrobend

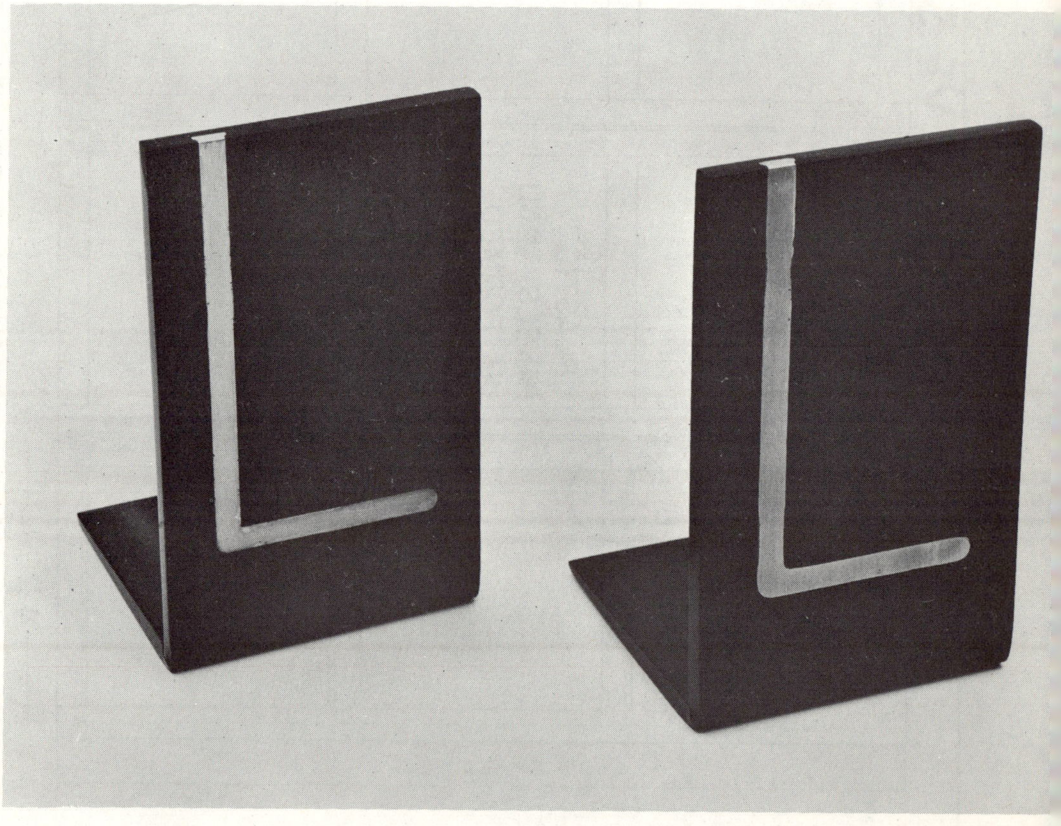

By Mervin Yue

BOOKENDS

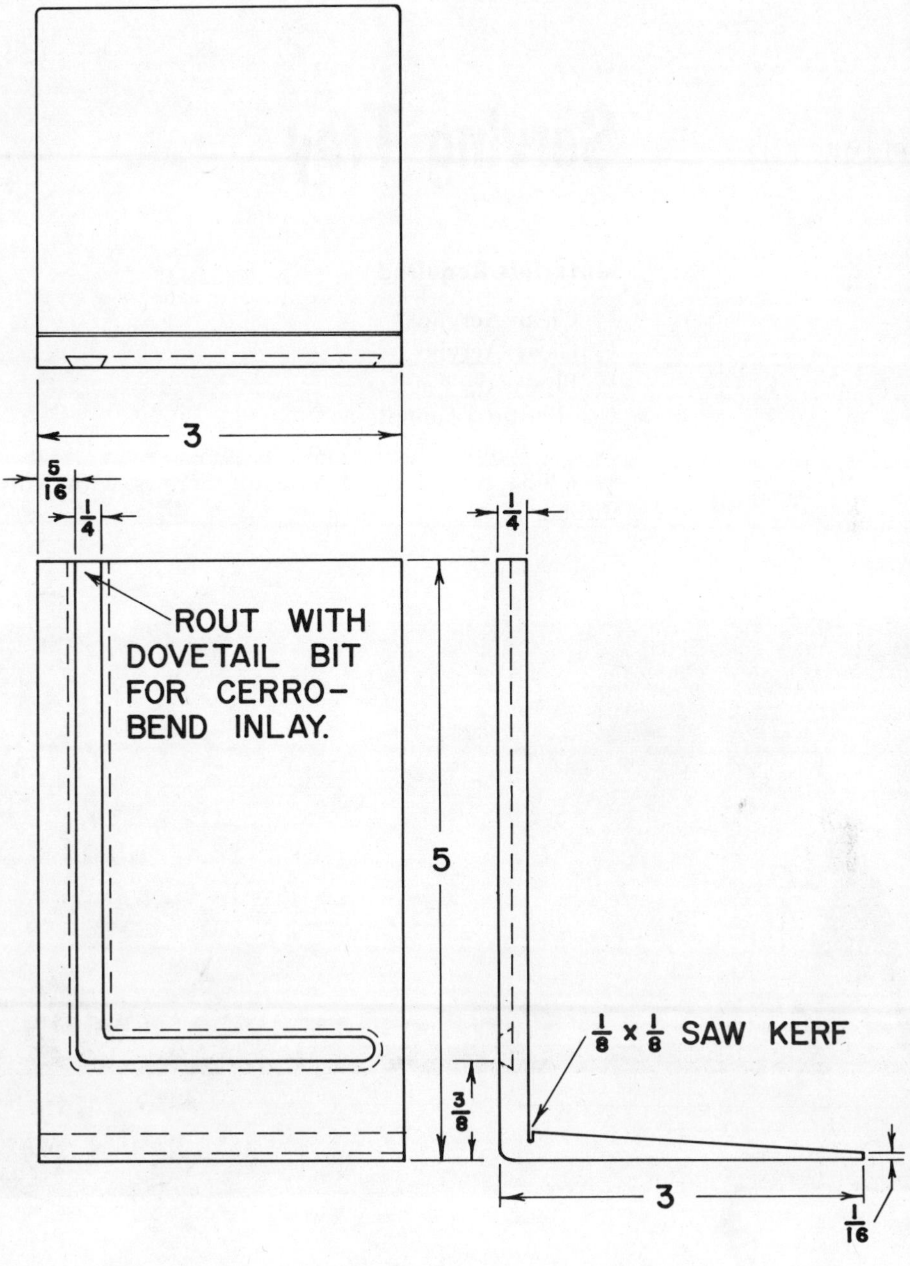

3
5/16
1/4
ROUT WITH DOVETAIL BIT FOR CERRO-BEND INLAY.
5
1/4
1/8 x 1/8 SAW KERF
3/8
3
1/16

PROJECT 42

Serving Tray

Materials Required

¼ Clear Acrylic
⅛ Clear Acrylic
¼ Plywood
1/16 Plastic Laminate
⅜ #6 Round Head Wood Screws
¾ #8-32 Round Head Machine Screws with Nuts
Rubber Tacks

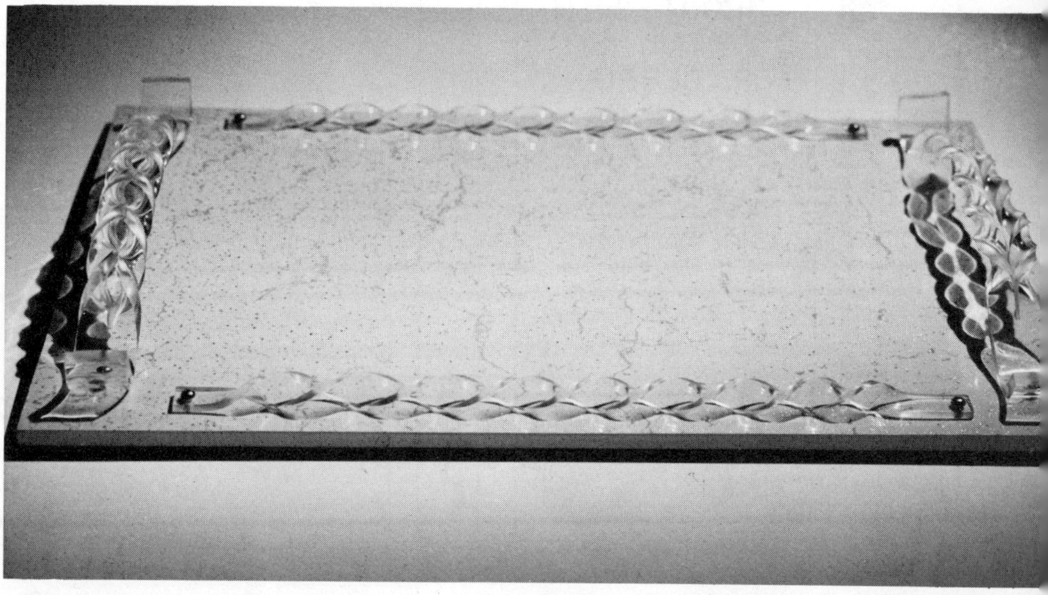

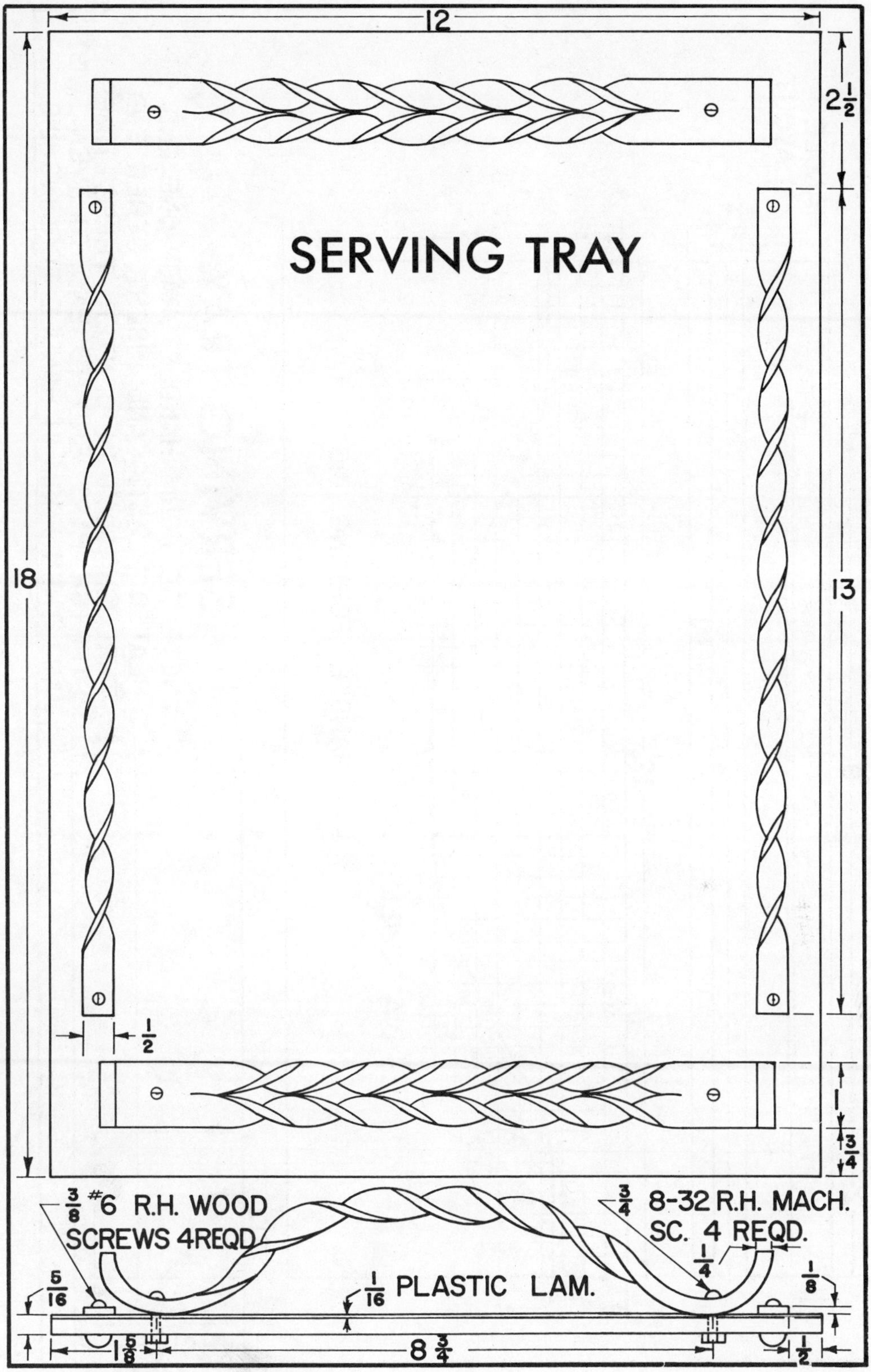
12
$2\frac{1}{2}$
SERVING TRAY
18
13
$\frac{1}{2}$
1
$\frac{3}{4}$
$\frac{3}{8}$ #6 R.H. WOOD SCREWS 4REQD
$\frac{3}{4}$ 8-32 R.H. MACH. SC. 4 REQD.
$\frac{1}{4}$
$\frac{5}{16}$
$\frac{1}{16}$ PLASTIC LAM.
$\frac{1}{8}$
$1\frac{5}{8}$
$8\frac{3}{4}$
$\frac{1}{2}$

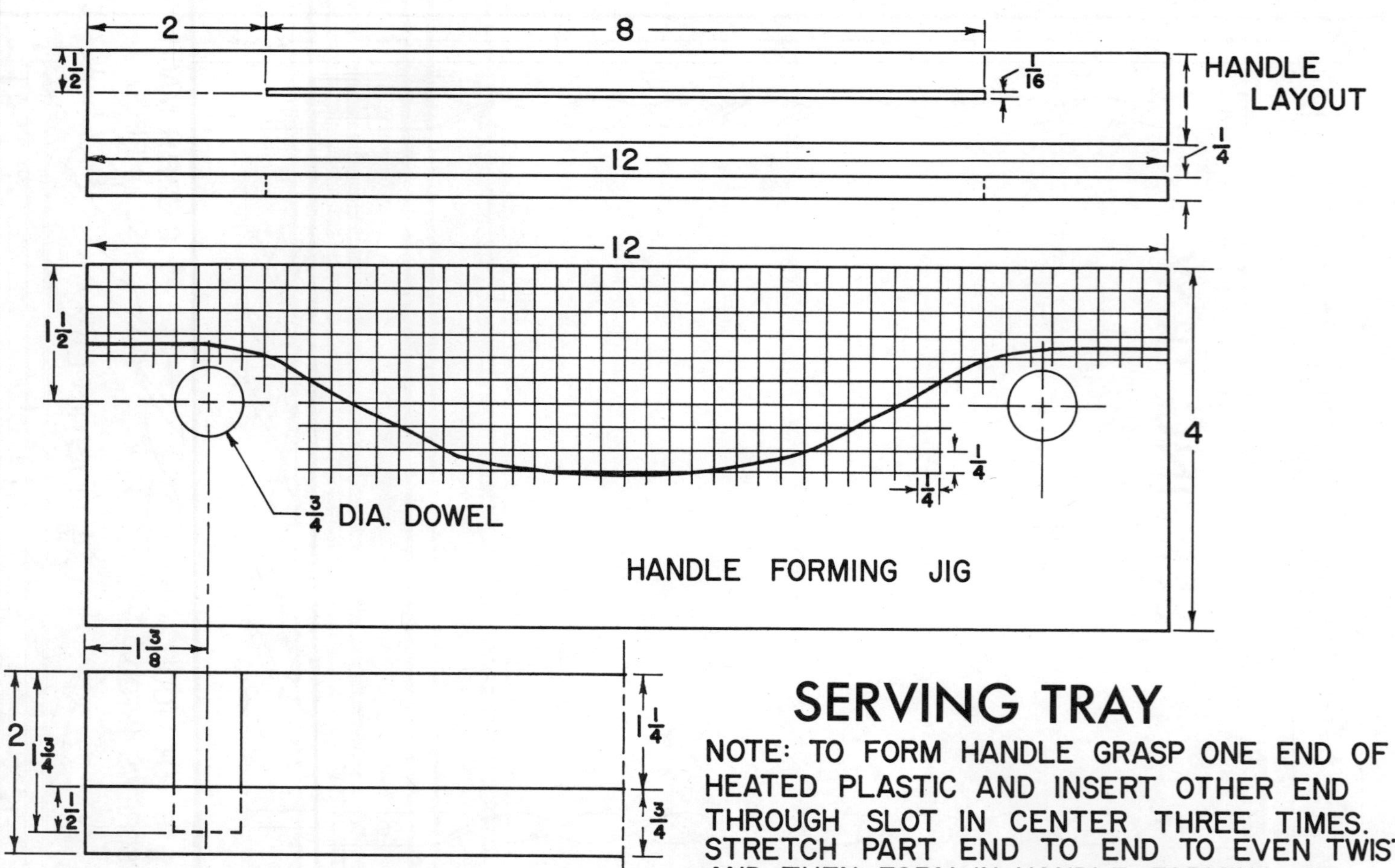

SERVING TRAY

NOTE: TO FORM HANDLE GRASP ONE END OF HEATED PLASTIC AND INSERT OTHER END THROUGH SLOT IN CENTER THREE TIMES. STRETCH PART END TO END TO EVEN TWISTS AND THEN FORM IN HANDLE FORMING JIG.

PROJECT 43

Paper Weight

Materials Required

Polyester Casting Resin
Hardener (Catalyst) for Resin

Refer to Technique 7 for information on casting polyester resin.

PROJECT 44

Pen Set

Materials Required

Polyester Casting Resin
Hardener (Catalyst) for Resin
Pen Funnels
Pen Swivel Bases
Pens

Refer to Technique 7 for information on casting polyester resin.

PEN SET

(POLYESTER RESIN CASTING)

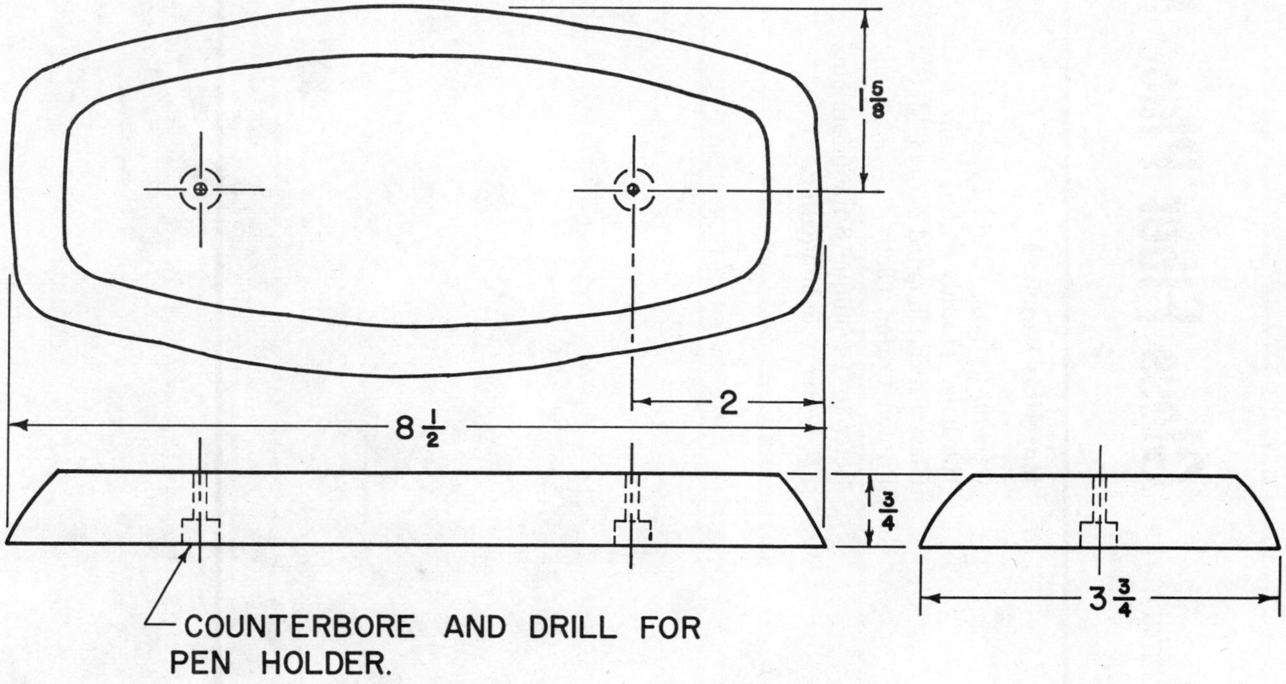

PROJECT 45

Glass Fiber Place Mats

Materials Required

General-Purpose, Rigid-Type Polyester Resin
Hardener (Catalyst) for Resin
Glass Fiber Chopped Strand Mat
Printed Cotton Cloth

Refer to Technique 8 for information on fiber glass-polyester resin laminating.

By Ronald Smith

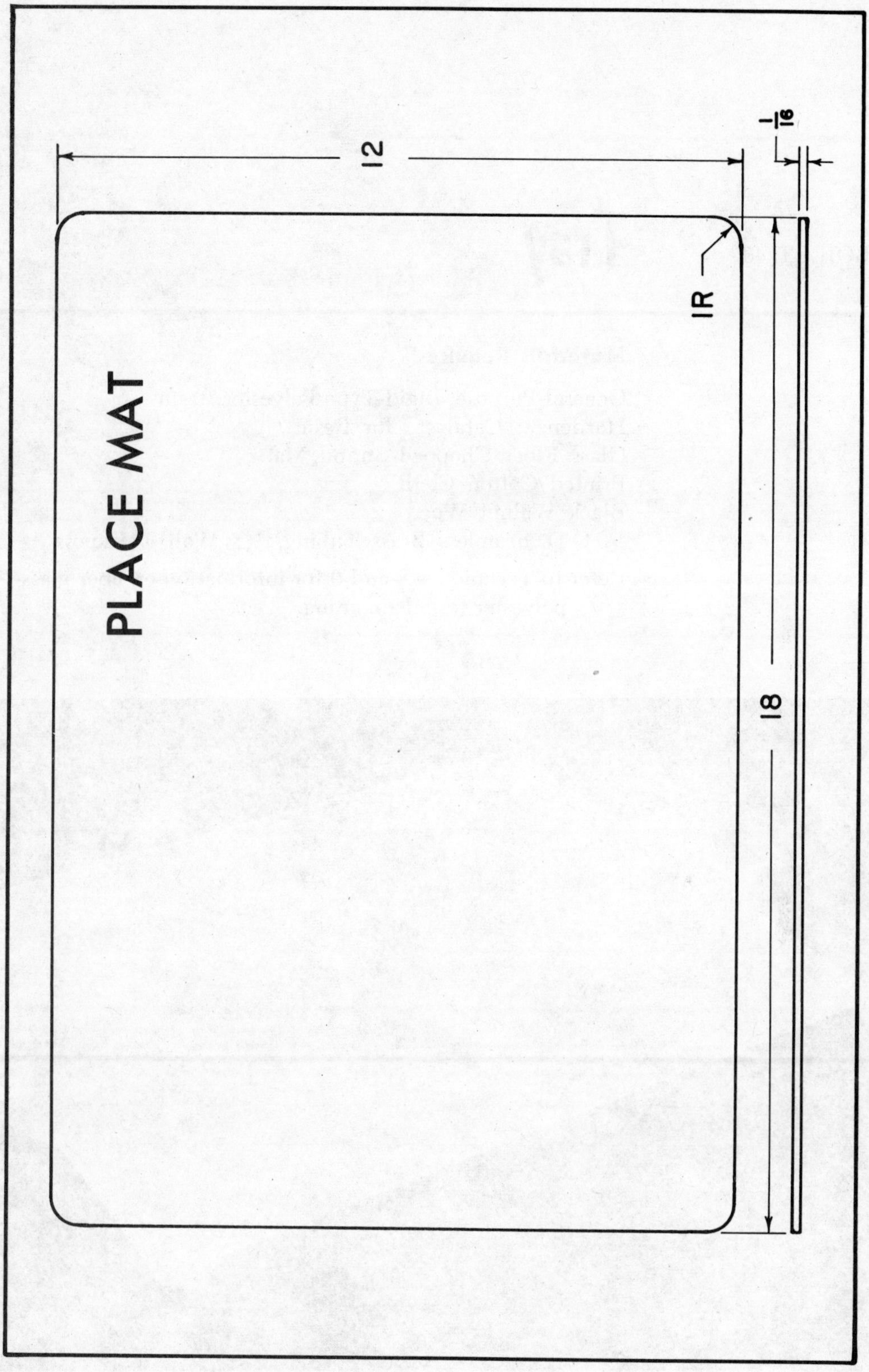
PLACE MAT
12
18
1R
1/16

PROJECT 46

Tray

Materials Required

General-Purpose, Rigid-Type Polyester Resin
Hardener (Catalyst) for Resin
Glass Fiber Chopped Strand Mat
Printed Cotton Cloth
Black Walnut Wood
½ O.D. Seamless Brass Tubing, $\frac{1}{16}$ Wall Thickness

Refer to Techniques 8 and 9 for information on fiber glass-polyester resin laminating.

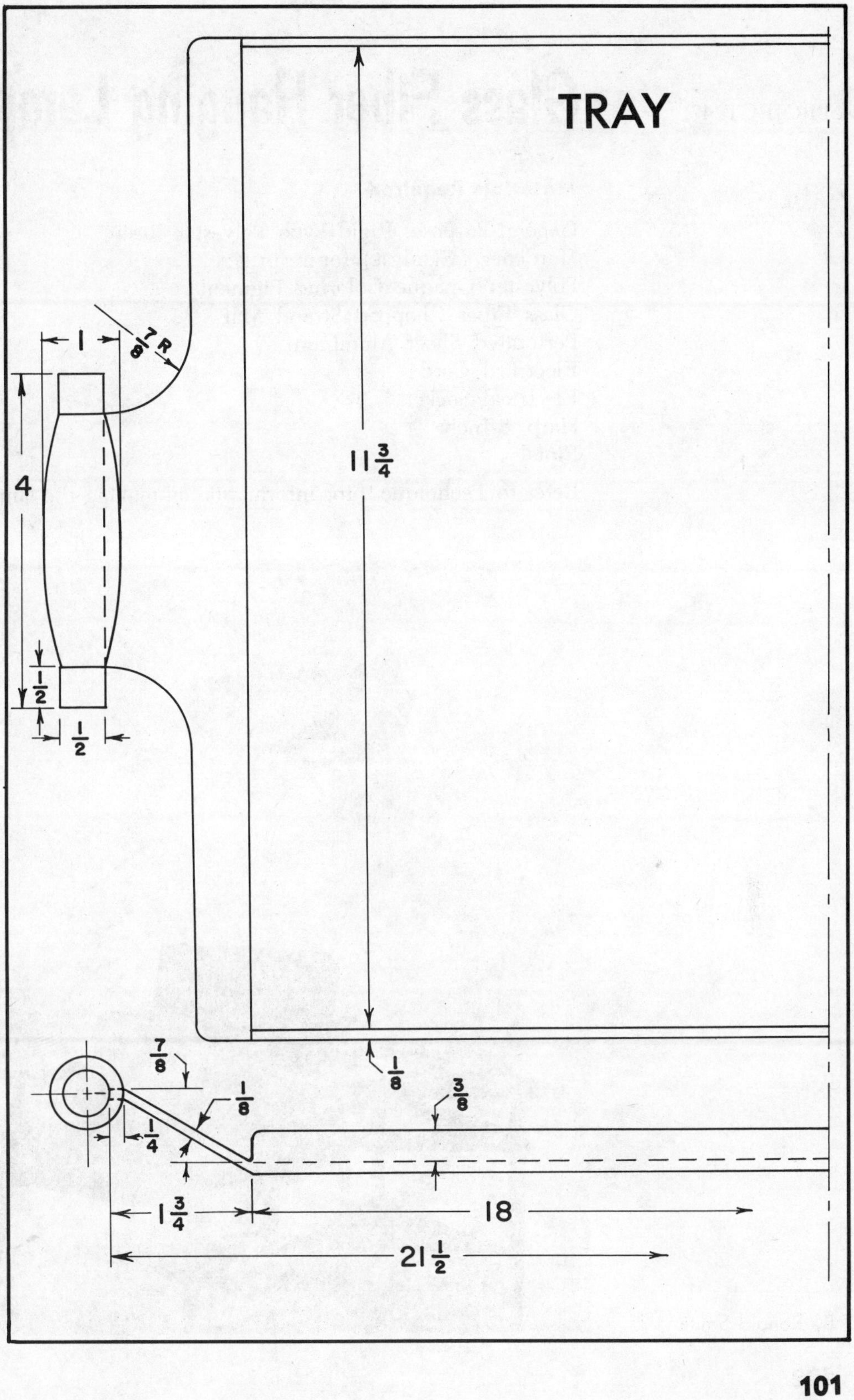
TRAY
1
7/8 R
4
1/2
1/2
11 3/4
1/8
7/8
1/8
3/8
1/4
1 3/4
18
21 1/2

PROJECT 47

Glass Fiber Hanging Lamp

Materials Required

General-Purpose, Rigid-Type Polyester Resin
Hardener (Catalyst) for resin
Polyester Opaque Coloring Pigment
Glass Fiber Chopped Strand Mat
Perforated Sheet Aluminum
Electrical Cord
Electrical Socket
Harp, 8-Inch
Finial

Refer to Technique 9 for information on making the lamp.

By Ronald Smith

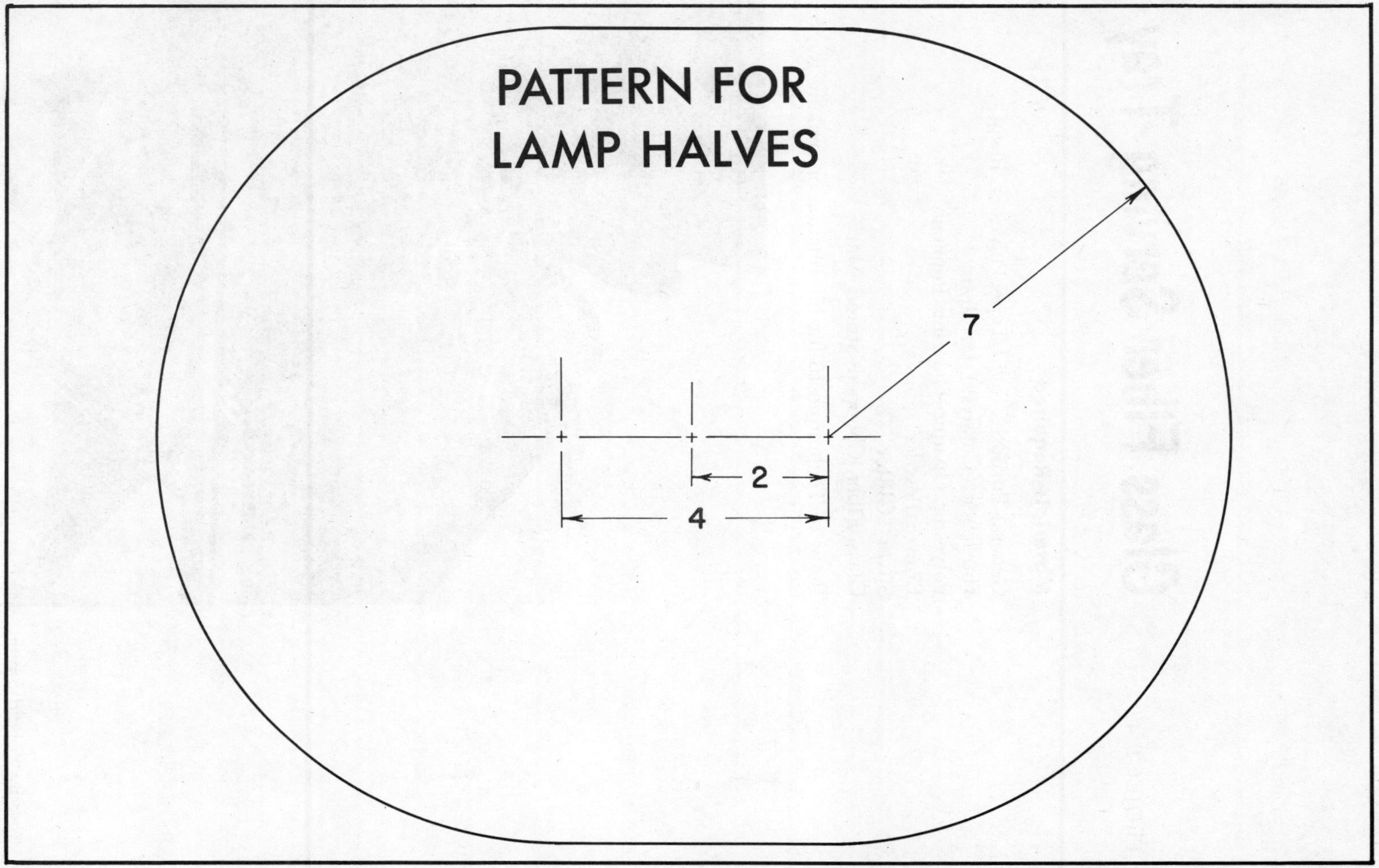
PATTERN FOR
LAMP HALVES
7
2
4

PROJECT 48

Glass Fiber Serving Tray

Materials Required

General-Purpose, Rigid-Type Polyester Resin
Hardener (Catalyst) for Resin
Polyester Opaque Coloring Pigment
Bronze Powder
Silver "Glitter"
Glass Fiber Chopped Strand Mat

Refer to Technique 10 for information on making the tray.

By Ronald Smith

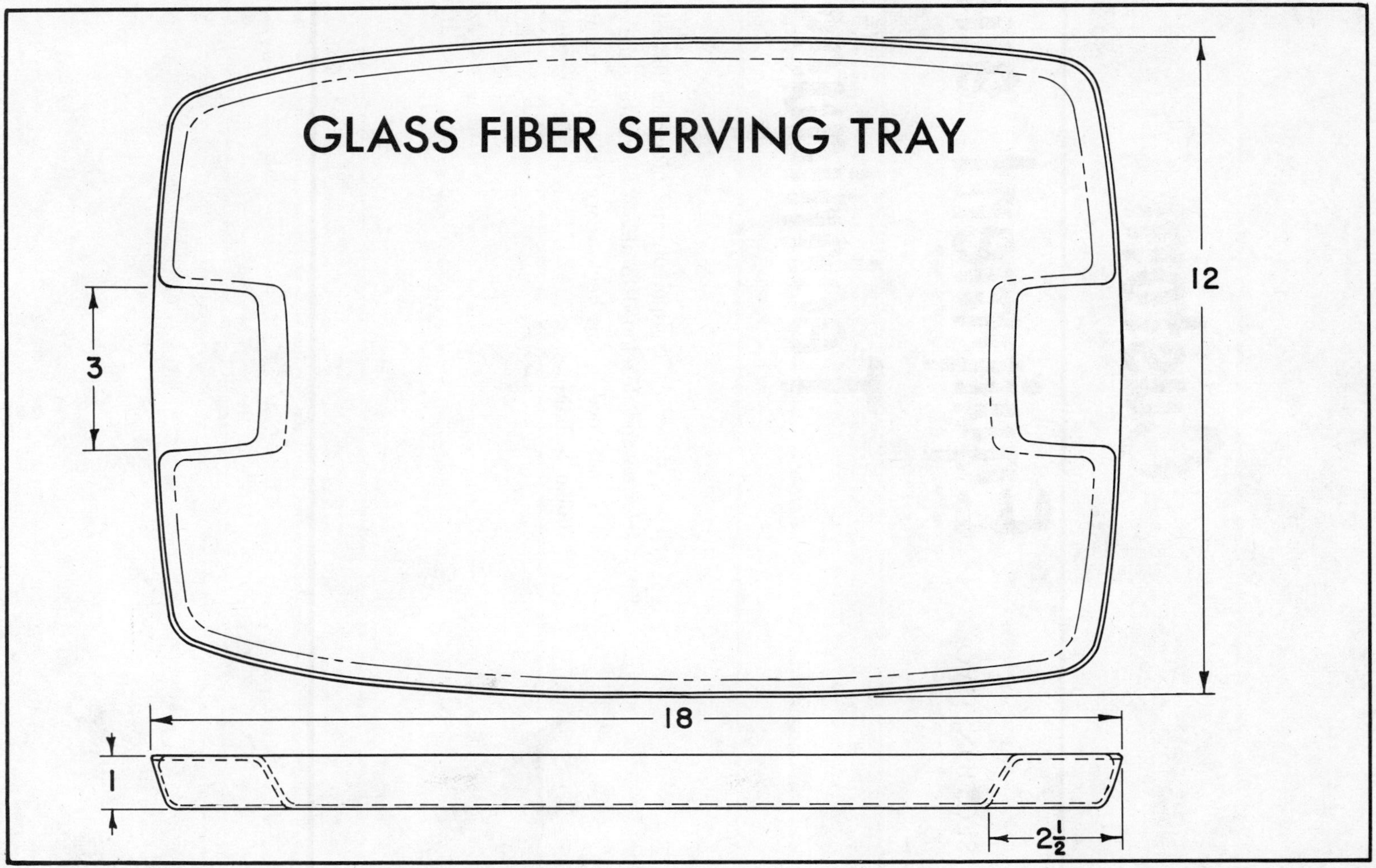
GLASS FIBER SERVING TRAY
12
3
18
1
2 1/2

SECTION TWO

Custom Equipment and Techniques

These materials, equipment, and processes have been selected and developed to aid in fabricating the projects contained in this book as well as others requiring similar techniques.

TECHNIQUE 1

Sandblaster

The sandblaster, Fig. 1, is constructed of ½″ plywood and is mounted upon a wood frame. The notations on the plan, Fig. 2, include the construction details necessary to make this piece of equipment. A clear acrylic plastic window is recommended since it is not only safer than glass, but may also be removed occasionally and buffed to eliminate the scratches caused by the impact of the flying particles of sand against it. Other recommendations for greater efficiency are:

A glass jar over a light bulb will protect it from flying sand and can be replaced when it has lost its transparency.

A rubber gasket, such as used in refrigerators, will seal the door against any leakage of dust.

Painting the inside of the unit white will increase the reflective value of the light.

An LA 7′ abrasive gun, (Fig. 3), a sharp-grained sand, and from thirty to fifty pounds of air pressure will give satisfactory results. Sandblasting produces a soft frosty appearance on plastic flowers. Complete opaqueness is possible if it is desired. With heavy paper or metal stencils securely fastened to the surface, designs, letters, and figures can be sandblasted on plastic. The sand is reusable. Dip the hopper of the gun to fill it as needed.

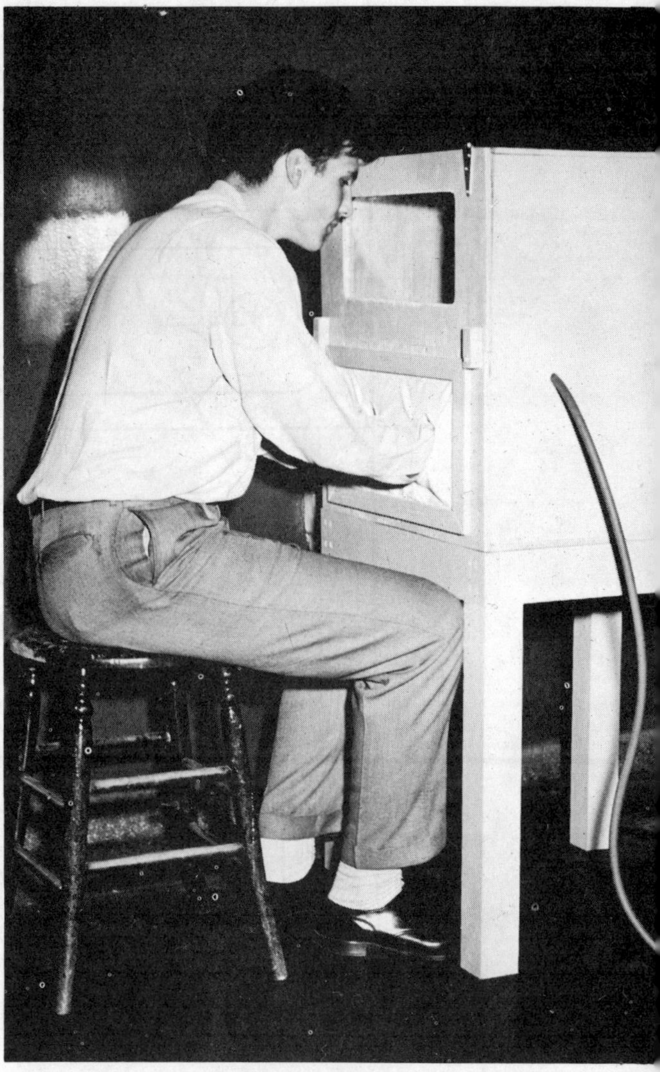

Fig. 1. Operating the Sandblaster

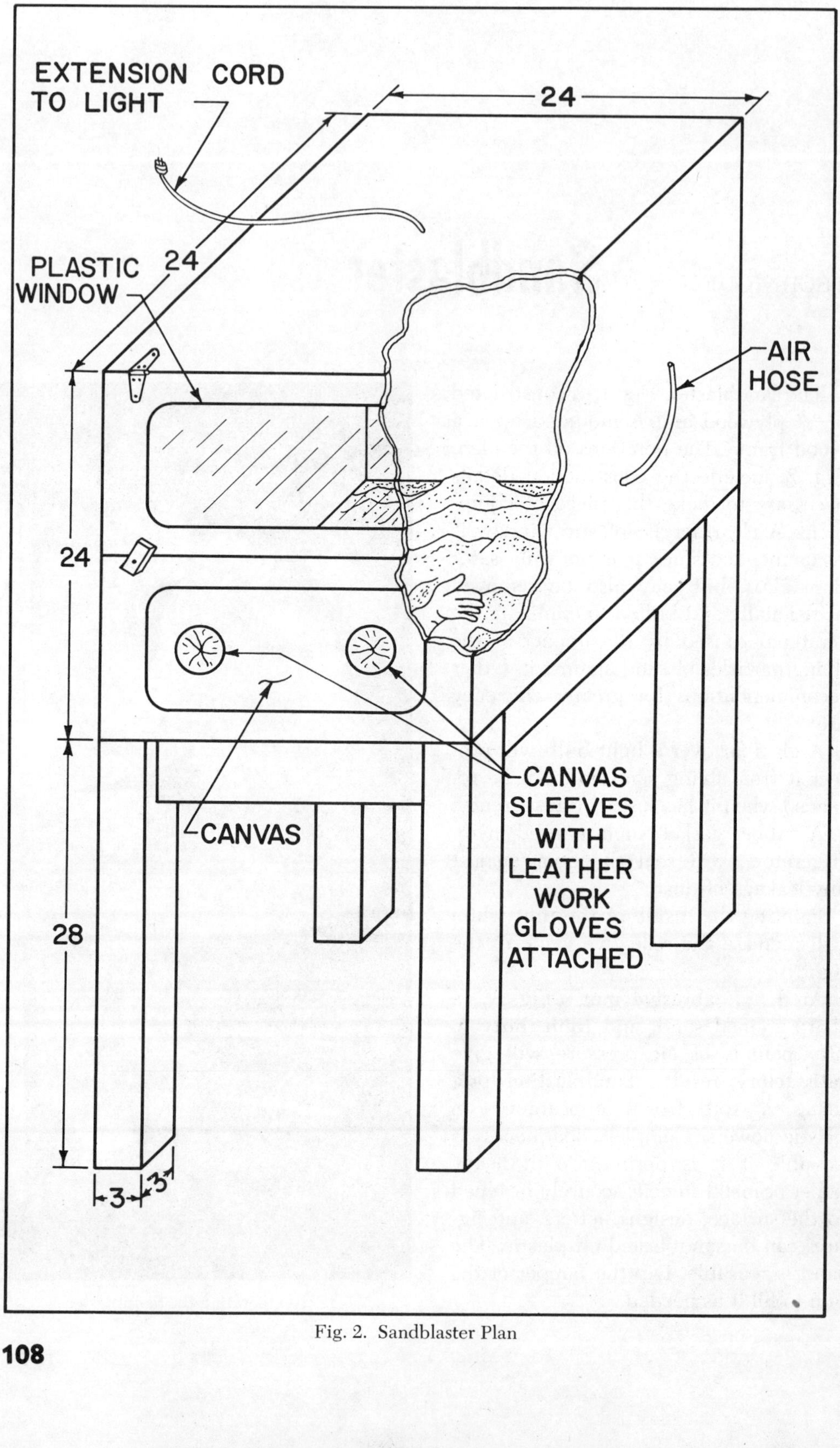

Fig. 2. Sandblaster Plan

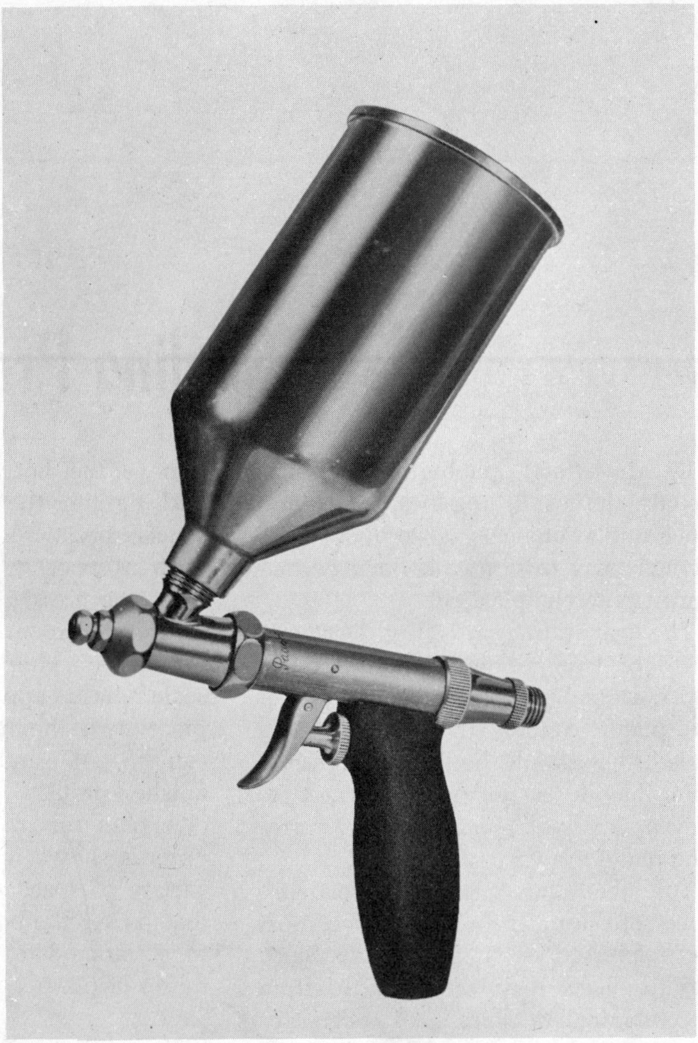

Fig. 3. LA 7′ Abrasive Gun (Paasche Airbrush Company)

TECHNIQUE 2

Heat-Sealing Press

Decal letters, photographs, coins, medals, charms, fishing flies, beads, and other similar objects (up to about $\frac{3}{32}''$ in thickness) can be embedded permanently in acrylic plastic.

The heat-sealing press, Fig. 1, is composed of two aluminum plates, bolts, and spacers. Since the inner surfaces of the plates contact the plastic being sealed, they should be lined with a material having a mirrorlike surface to prevent "mark off" on the plastic. Ferrotype plates are very satisfactory.

To embed materials, the two or more pieces of plastic to be heat-sealed must be compressed by $\frac{1}{16}''$ while hot. Since acrylic sheets may vary a little from their nominal thickness, it is necessary to measure the thickness carefully after the masking paper has been removed. With $\frac{5}{16}''$ spacers between the plates, the total thickness of the plastic should be $\frac{3}{8}''$, not including the object to be embedded. Any combination of thicknesses totaling $\frac{3}{8}''$ may be used.

After all surfaces of the plastic are cleaned, the sandwich, including the parts to be embedded, is held together with cellulose tape, and lightly clamped in the press. This is placed in an oven for one hour at 275° F., or until the plastic is rubber-like when pressed with a pointed instrument.

Protect your hands with gloves. Then remove the hot press from the oven and tighten the nuts, as far as the spacers permit. Return the press to the oven at once. Shut the heat off and leave the press in the oven until it has returned to room temperature.

When unclamped, the sheets of heat-sealed plastic appear to be a single solid piece with the material embedded. It can be cut, sanded, and buffed to a finished project.

A plan for making the heat-sealing press is shown in Fig. 2. It includes a variety of spacers which may be used to slip over the bolts between the plates to accommodate any of several thicknesses of plastic.

Fig. 1. Heat-Sealing Press
Note the metal plates, bolts, nuts, washers, spacers, and burnisher for cleaning plastic before sealing.

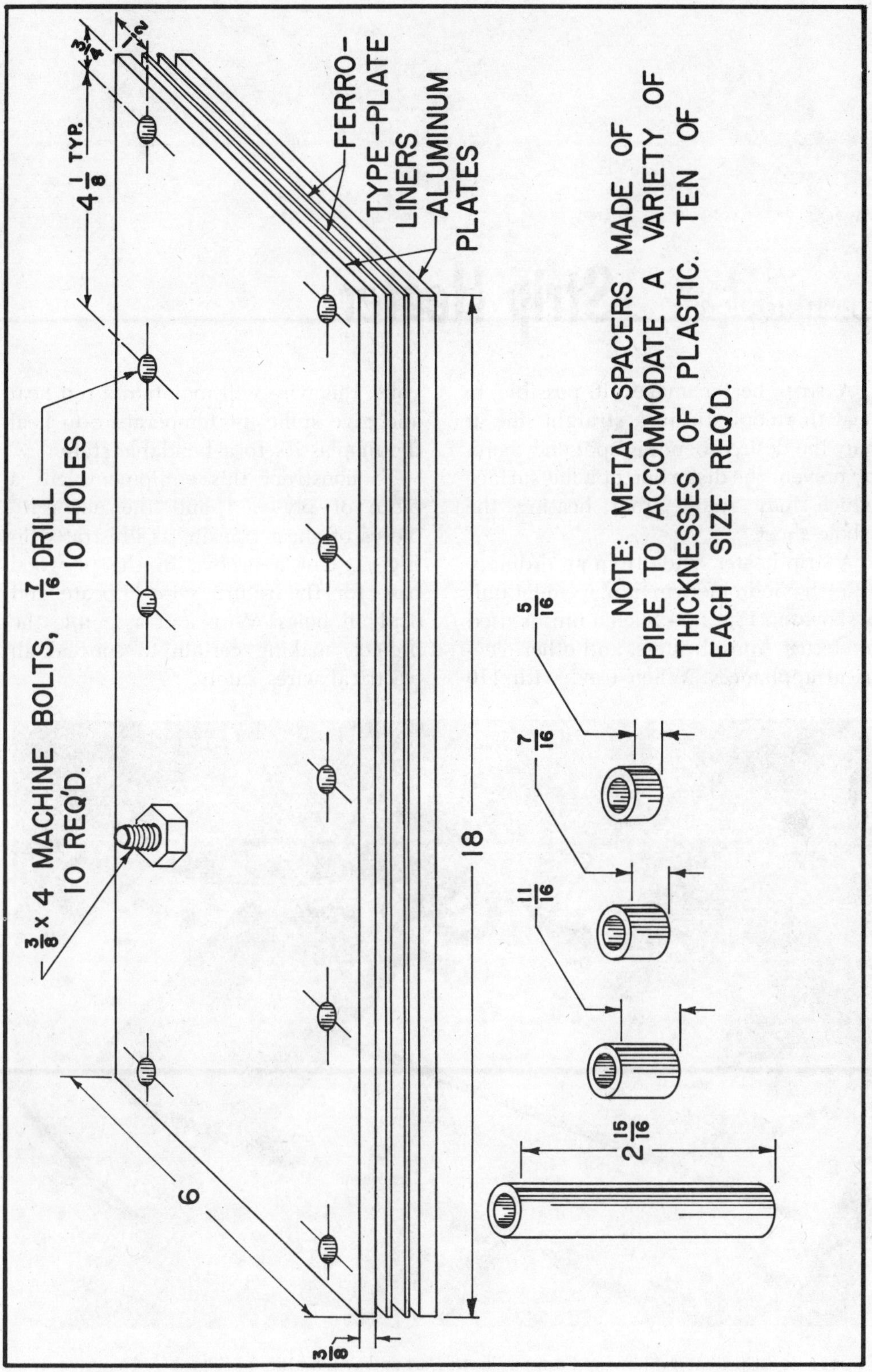

Fig. 2. Plan for Heat-Sealing Press

TECHNIQUE 3

Strip Heater

A strip heater makes it possible to heat thermoplastic in a straight line at only the desired bending spot and thereby prevent the distortion of a flat surface which may result from heating the whole sheet.

A strip heater made from an ordinary 12-inch Nichrome wire replacement unit is shown in Fig. 1. – Such a unit is used in electric irons, heaters, and other electrical appliances. When used with 110 volts, this wire will maintain a red heat and give sufficient temperature to heat thermoplastics to a bendable state.

To construct this equipment, cut a piece of plywood and the necessary pieces of sheet transite as illustrated in Fig. 2. Cut a groove in the plywood base for the return wire. Locate and drill all holes. Wire and assemble the project, making certain to conceal all electrical wires safely.

Fig. 1. Strip Heater, Showing Formed Plastic Part and Second Piece in Place for Heating

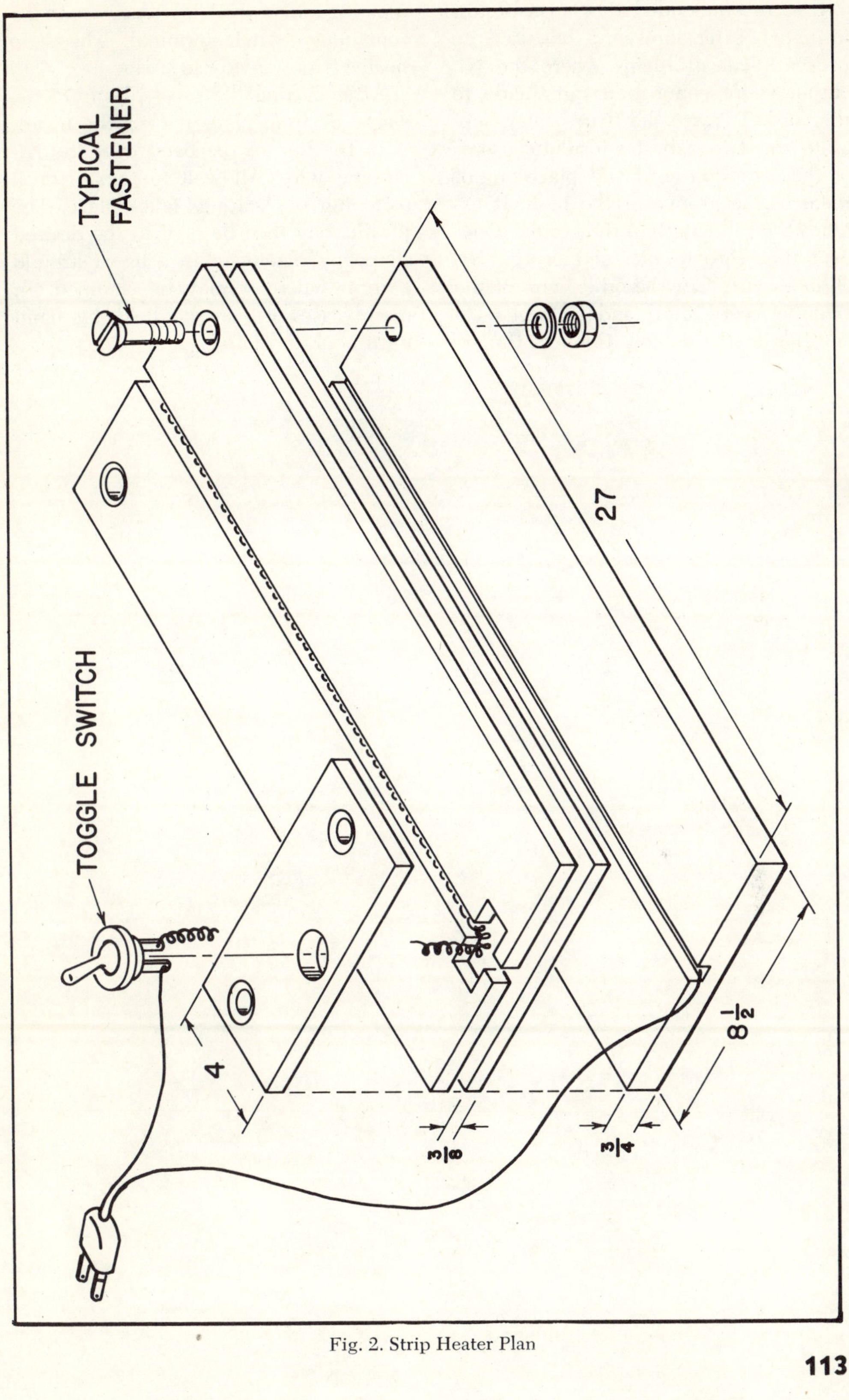

Fig. 2. Strip Heater Plan

It is recommended that an asbestos-insulated extension cord be used to prevent heat problems where the Nichrome wire connections are made to the cord. To wire the strip heater, separate the two asbestos-insulated wires of the extension cord and place one of them in the groove of the base. It extends the full length of the groove. Connect this wire to one end of the Nichrome wire. Cut the other wire of the extension cord short and connect it to one side of the switch. Connect the remaining end of the Nichrome wire to the remaining switch terminal. The strip heater is now ready to use.

After cutting the sheet plastic to the shape desired, place it over the heater with the line of the bend over the Nichrome wire. Allow it to remain until softening or "wilting" takes place. The plastic may then be bent to the desired shape by placing it in a jig or around a form where it may be clamped. In many cases it may be held by hand until cool and hard.

TECHNIQUE 4

Air Pressure Forming Equipment

Thermoplastic bowls and covered dishes are two of the many objects which can be made from parts that are blown, or formed with air pressure. Only simple equipment is required, Fig. 2. It can be easily constructed from ¾" plywood, toggle clamps, some rubber gasket material, and a piece of sheet metal.

The plywood base upon which the toggle clamps are mounted is the permanent part of the equipment. See Fig. 1. In the center of the base is an opening leading from a source of constant air pressure. At least ten pounds of pressure is necessary. A sheet of rubber gasket material, with a hole at the center to allow the compressed air to pass through, is placed upon the base. This gasket forms an airtight seal between the plastic and the base as the plastic is blown. A shut-off valve controls the flow of air. The air inlet is covered by a piece of sheet metal, raised above the hole by small washers and screws. This deflector prevents the incoming rush of cold air from cooling the plastic unevenly as blowing takes place. The shaped die, shown in Fig. 1 being held

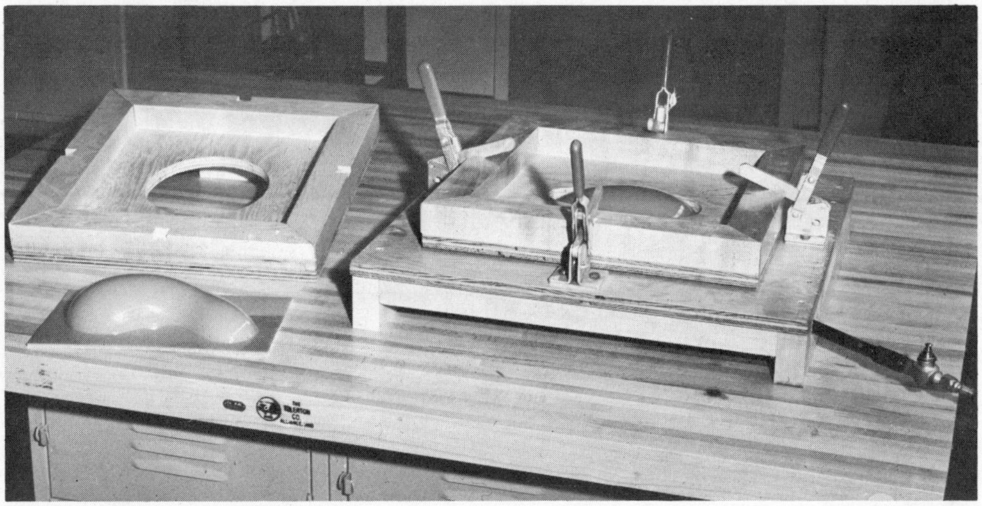

Fig. 1. Air Pressure Forming Equipment in Operation with Blown Part and Second Form Adjacent

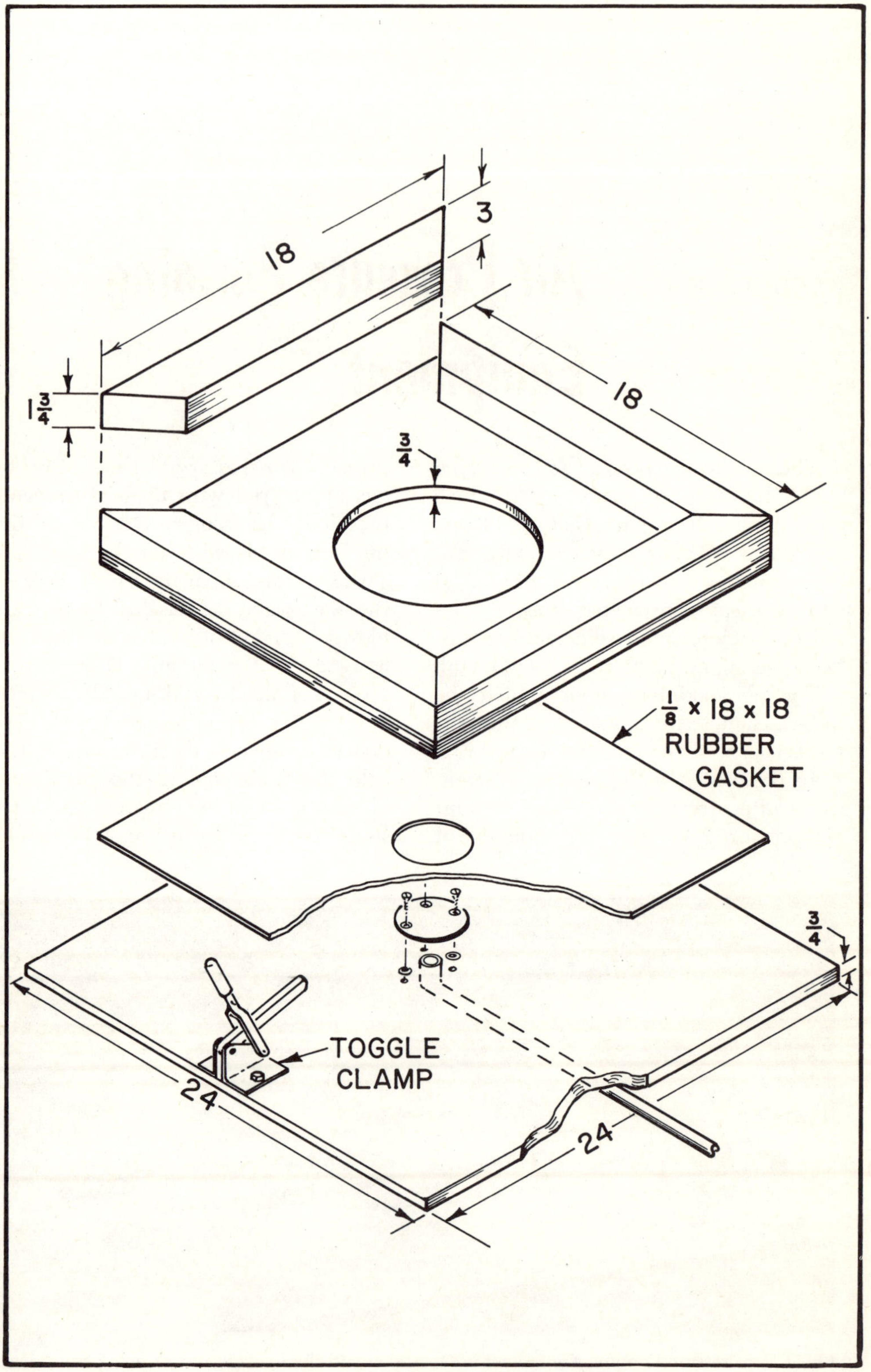

Fig. 2. Plan for Air Pressure Forming Equipment

in place by the toggle clamps, is removable. The size and shape of the hole in this part determines the size and outline of the formed plastic.

Since plastic stretches and thins out as it is blown, a piece ⅛″ thick is necessary to blow a hemisphere six inches in diameter. If larger dimensions are required, then the thickness of plastic must be increased accordingly.

To use the pressure forming equipment it is necessary cut to the plastic so that at least a one-inch margin is allowed around the opening in the die through which the material is to be blown. Thus, for a die six inches in diameter, an eight-inch square piece of plastic ⅛″ thick is needed. Open the toggle clamps, remove the die, and regulate the air pressure to ten pounds. Heat the plastic until it is rubber-like, then remove it from the oven and center the piece over the air inlet. Center the die over the plastic and clamp it in place. Then open the shut-off valve and allow the air to exert pressure under the plastic, thus blowing it up through the hole to form the desired shape. Close the valve.

When the blown part has cooled, it is removed and the excess material cut off. Sand and buff to finish the edge.

TECHNIQUE 5

Vacuum Forming Equipment

Wall plaques and decorative screens are two of the many objects which can be made from parts that are vacuum formed.

The equipment in Fig. 1 is constructed from ½″ and ¾″ plywood, 1″ dowel, ¾″ pipe, a pipe "T", a winged stud, a pipe flange, metal window screen, sheet steel, a suitcase fastener, hinges, rubber gasket material, and a heat lamp.

Construction Procedure

1. Make the base.
2. Cut the pieces for part "A".
3. Bore a hole in one side in which to thread a piece of pipe. This pipe will lead to a source of vacuum. An ordinary vacuum cleaner generates sufficient suction.
4. Assemble the four sides of part "A" with airtight joints; seal and fasten this assembly to the base with wood screws.
5. Cut a piece of metal window screen to fit the top of part "A" and tack it in place.
6. Cut a rubber gasket to fit around the top edge of part "A" and cement it in place.
7. Next, cut the pieces for part "B".
8. Cut out one side to fit down over the pipe in part "A".
9. Bore holes halfway through the two opposite sides to take the dowel handles. Fasten the handles in place with glue and wood screws.
10. Assemble the four sides of part "B" with airtight corner joints, and with just enough clearance to slip down over part "A".
11. Lay out and cut a piece of sheet steel to frame the top of part "B".
12. Fasten the metal frame in place with hinges and a suitcase fastener.
13. Drill and tap threads in the pipe "T" to take the winged stud. This will allow the heat lamp to be adjusted to the proper height above the plastic for efficient heating.
14. Assemble the pipe lamp support parts, align the heat lamp directly over part "B" and fasten the pipe flange to the base to complete the equipment.

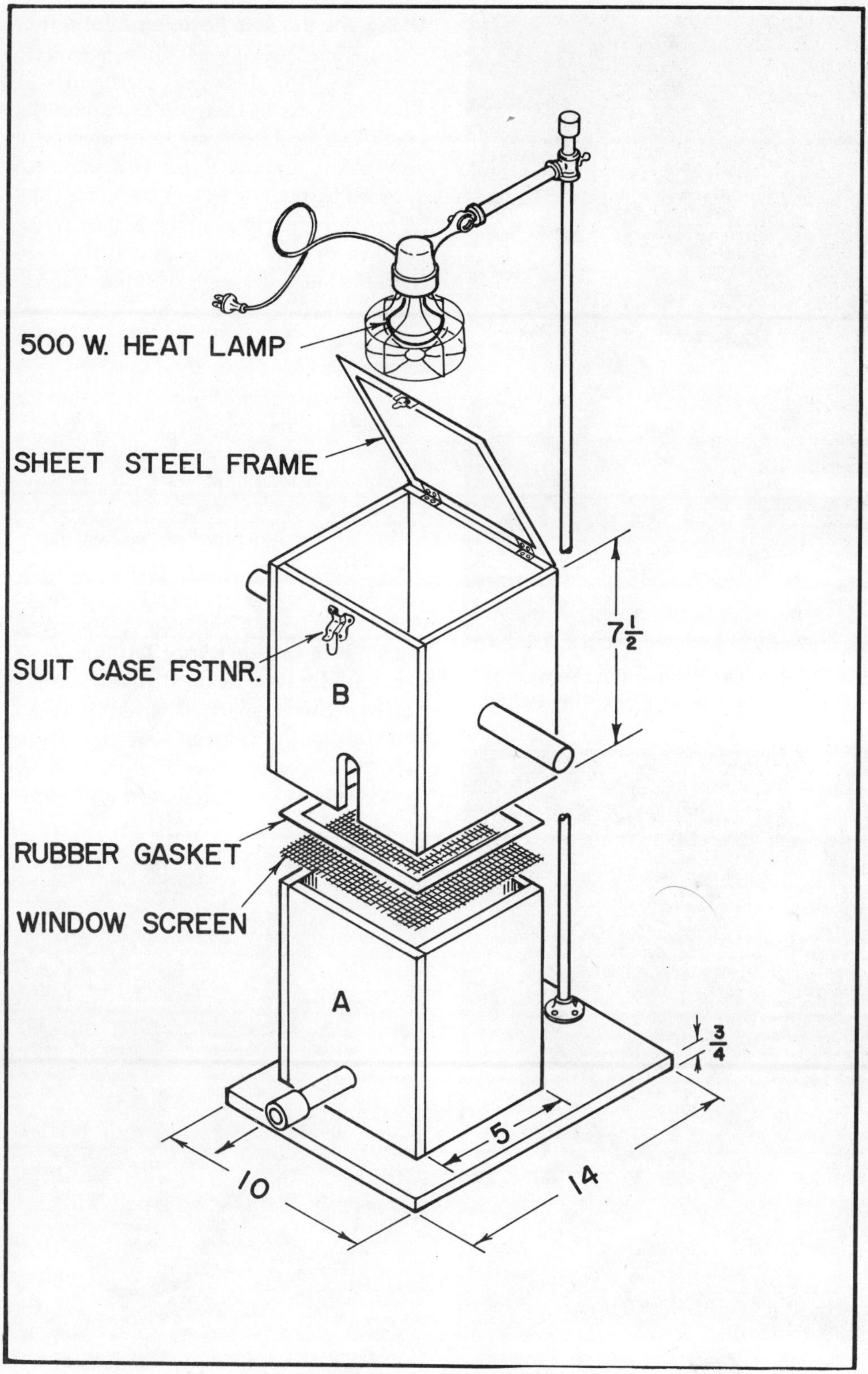

Fig. 1. Plan for Vacuum Forming Equipment

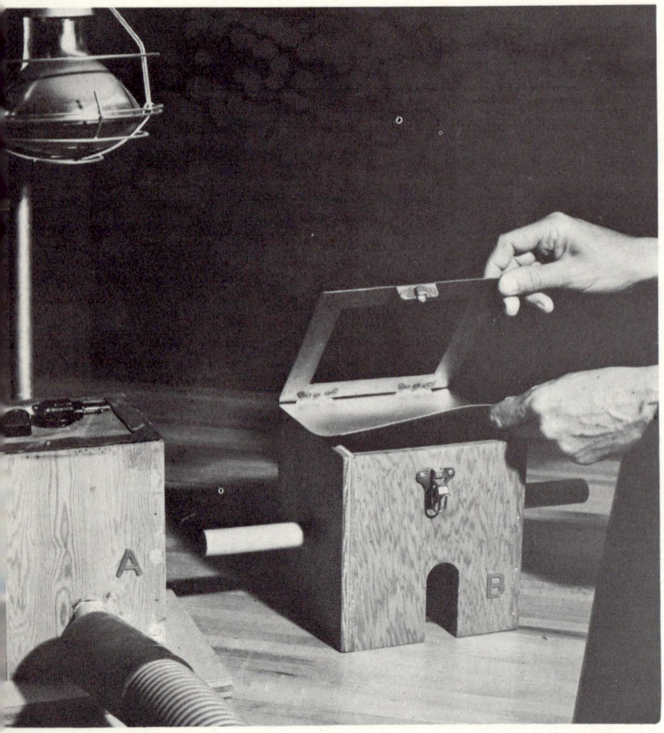

Fig. 2. Model Positioned on Part A
and Plastic Sheet Being Clamped in Part B

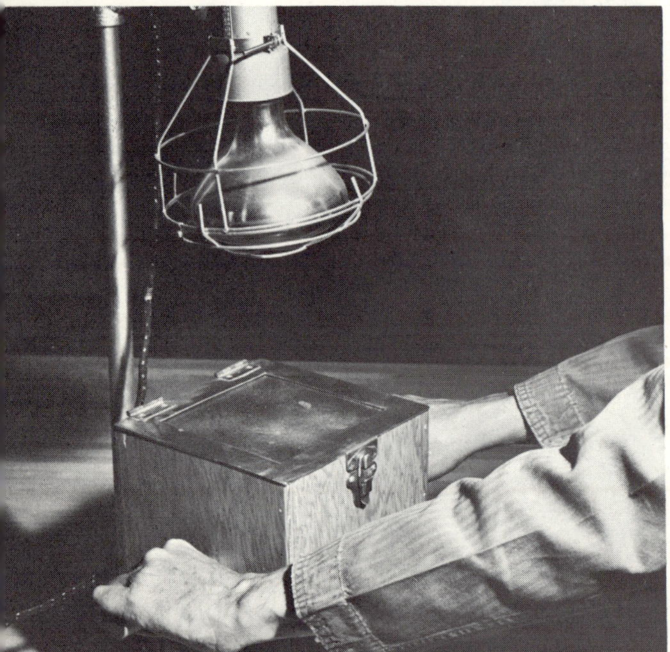

Fig. 3. Heating Plastic Sheet

Using the Vacuum Forming Equipment

1. Select the model upon which to form the .010 thickness thermoplastic. This thickness of thermoplastic material (cellulose acetate) can be drawn with fine detail, and without thinning out appreciably, over a model not more than ⅜″ in thickness. The model may be coins, keys, medals, or even plants, if they are rigid enough to support the plastic sheet as it is heated and vacuum drawn.

2. Place the model on the window screen, part "A", Fig. 2.

3. Clamp a sheet of plastic under the metal frame of part "B", Fig. 2.

4. Slip part "B" over part "A", but do not touch the model with the plastic, Fig. 3.

5. Adjust the height of the heat lamp above the plastic. Experiment until the heat lamp causes the plastic to change from a glossy to a dull appearance and sag slightly in a matter of seconds.

6. At the very moment that the plastic is hot, lower part "B" over part "A" and turn on the vacuum.

7. The formed plastic part, Fig. 4, may be immediately removed and used.

Fig. 4. Vacuum Formed Plastic Part

TECHNIQUE 6

Forms for Wood and Plastic Resin Laminating

The forms in Fig. 1 were constructed of wood and used to form the clothes hanger, Project 34. Other projects, such as water skiis and archery bows, may be laminated using the same basic principle for developing the required forms.

The clothes hanger was designed by bending a piece of sheet metal until the desired shape was obtained. This pattern was used to lay out the forms. It is very important to allow the exact amount of clearance between forms "A" and "B" for the thickness of the wood veneers to be used. See Fig. 3. Satisfactory results were obtained by using ⅛″ poplar core and 1/28″ mahogany veneers for the clothes hanger.

Lay out the forms, saw to shape, and sand smooth. Attach guides to insure accurate alignment of "A" and "B" when using the forms. A coat of oil or wax will prevent any excess plastic resin from sticking to the forms.

Making a Clothes Hanger

To make a clothes hanger, cut one piece of poplar ⅛″ x 2″ x 20″ and two pieces of mahogany veneer 1/28″ x 2″ x 20″. Any one of several thermo or thermosetting plastic resin adhesives may be used to bond these pieces together. Coat both sides of the poplar core stock and one side of each piece of mahogany veneer. Match all three pieces together and center them upon form "A", Fig. 2. Two small brads driven through the pieces near the outside edges and into the form will keep them from slipping. Next, place forms "A" and "B" together as shown in Fig. 3, and clamp them securely.

When the laminated piece is thoroughly cured, remove it from the forms, develop the pattern for the hanger and trace it upon the lamination. Saw and sand the part to shape. Drill a hole at the center of the hanger, and make a hook from acrylic rod stock 5/16″ diameter by 7″ long. Insert the hook in the hanger and cement a ⅛″ x ½″ dia. piece of acrylic to the end of the plastic hook to prevent it from pulling through the wooden hanger during use over a period of time.

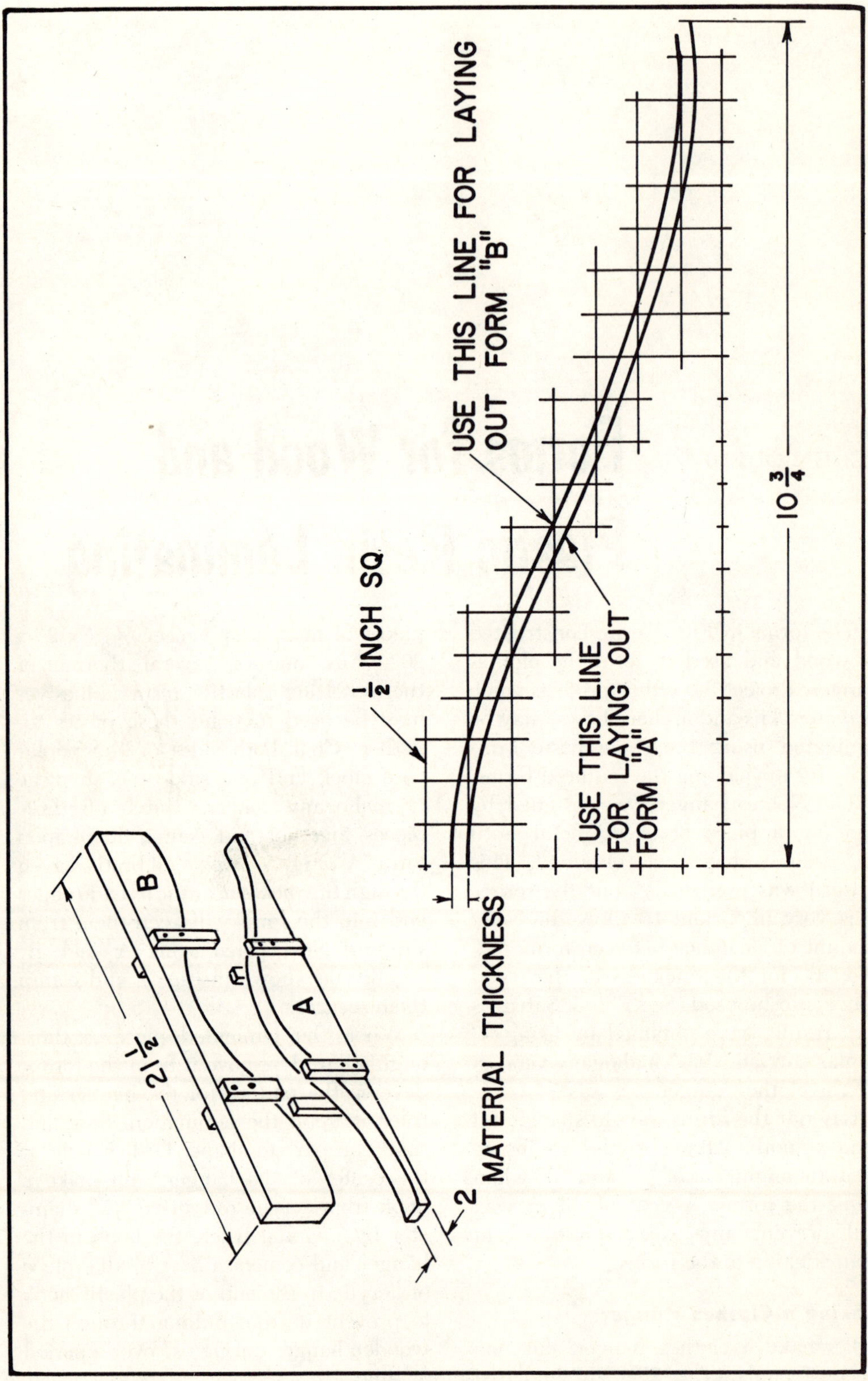

Fig. 1. Layout for Constructing Wood Forms

Fig. 2. Adhesive Coated Parts Fastened in Place for Clamping

Fig. 3. Pressure Applied to Forms

TECHNIQUE 7

Casting Polyester Resin

Figs. 1 through 6 illustrate the simple steps in embedding a flower in polyester casting resin. For best results when embedding a flower, or any other organic material, the specimen should be thoroughly dried. This can be accomplished by placing the freshly cut flower in a box between layers of dry sand, and then placing the box in a warm dry environment (such as on top of an automatic hot water heater) for several days.

Secure the container to be used for the mold. Inexpensive glass dishes make ideal molds. Coat the container with mold release. Determine the amount of resin needed and measure one-half of the resin to be used into the mixing dish, Fig. 1. Use a tablespoon. Fig. 2 shows adding the hardener (catalyst) three drops per tablespoonful of resin. Mix the hardener into the resin thoroughly, but stir slowly to prevent air bubbles from accumulating in the mixture. Then pour the resin into the glass or metal mold which has been coated with a mold release agent, Fig. 3.

Allow the mixture in the mold to jell, then mix the remaining half of the resin with the hardener. Now pick up the flower with tweezers and submerge it in the freshly mixed batch of resin to coat it with resin. Remove the flower and place it upon the jelled resin in the mold, Fig. 4. Pour the remaining resin into the mold, Fig. 5. Any bubbles which appear can be drawn to the surface with a broomstraw, wire, or long needle. Equipment may be cleaned with lacquer thinner or acetone. Allow the casting to harden, then remove it from the mold, Fig. 6. It is then ready for sawing, sanding, buffing, or other machining and finishing operations.

This process may be used for embedding photographs, insects, sea shells, medals, and many other objects. The resin may be used in its clear form or colored. It can be cast into molds to form many art objects.

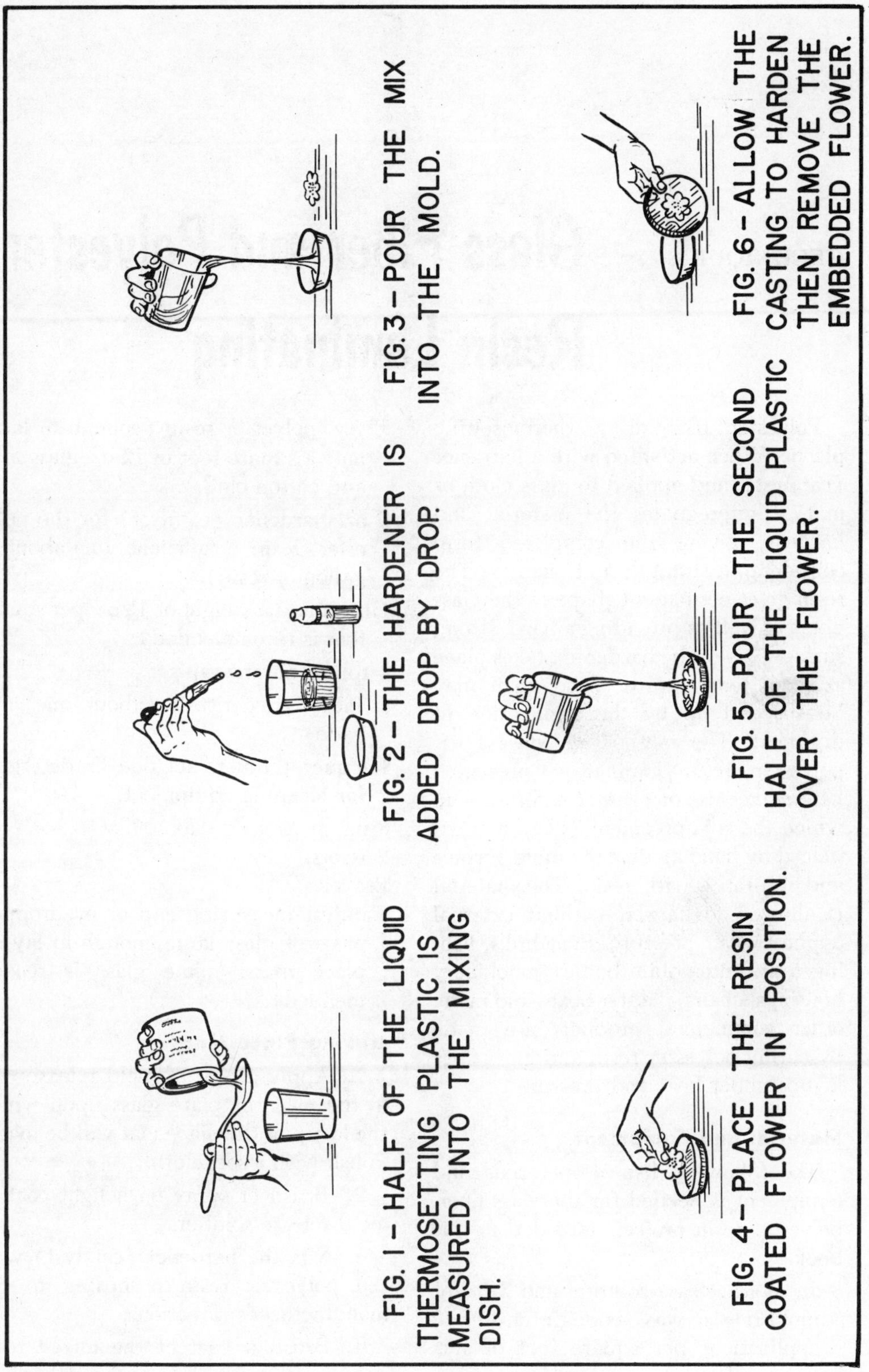

Fig. 1. Steps in Casting

TECHNIQUE 8

Glass Fiber and Polyester Resin Laminating

Polyester resin is a thermosetting plastic. When activated with a hardener (catalyst), and applied to glass cloth or mat, it impregnates the material and hardens, giving the completed form considerable strength and rigidity. The resin gives permanent shape to the glass fibers as well as providing a hard, strong surface finish. The hardening takes place at room temperature, although it may be speeded up by the application of dry heat. The method of glass fiber, polyester resin laminating presented here makes use of *contact molding* — in which the reinforcement (glass mat) is placed by hand against the mold surface and saturated with resin. The material is allowed to harden without external application of pressure. Boat hulls, fishing rods, automobile bodies, machinery housings, chairs, planter boxes and many other commercial products are made by curing polyester resin and glass fiber layups under heat and pressure .

Material and Equipment

The following list of materials and equipment is needed for the glass fiber, polyester resin projects included in this book.

8 oz. mold release solution and 2 oz. of mold release wax (enough for 8 or 10 applications per square foot of surface).

12 oz. polyester resin (enough to laminate a square foot of 1½ oz. glass mat, and cotton cloth).

½ oz. hardener (catalyst) for the polyester resin (sufficient for about ½ gallon of resin).

Glass mat (a weight of 1½ oz. per square foot is recommended).

1 roll of Saranwrap.

1 inexpensive brush (about one inch wide).

Lacquer thinner, acetone or detergent for cleaning equipment.

8 oz. measuring cup.

Scissors.

Brayer.

Catalyst measuring cup or eyedropper.

1 piece of glass large enough to lay up place mats (plate glass is recommended).

Making Place Mats

1. Apply a coat of mold release wax to the piece of plate glass upon which the layup of the place mat will be made. Polish with a soft cloth.

2. Brush or spray on a light coat of mold release solution.

3. Mix the hardener (catalyst) with the polyester resin according to the manufacturer's directions.

4. Brush a coat of the mixed resin over the waxed surface of the plate glass.

5. Place the printed cotton cloth (finished side down) upon the coat of resin and coat it with more resin.

6. Place a layer of glass mat on top of the cotton cloth.

7. Dab the glass mat with the brush until all fibers are saturated.

8. Place Saranwrap over the layup; roll out air and smooth the surface by working from the center to the outside with the brayer.

9. After the laminate has cured, remove the Saranwrap.

10. Cut the piece to shape using a fine-toothed metal-cutting saw.

11. Dry-sand and then wet-sand to smooth the edges and complete the place mat. (Project 45).

TECHNIQUE 9

Sheet Metal and Plywood Mold Making

Many glass fiber projects composed of simple curves may be formed on molds of sheet steel and plywood. Fig. 1 shows the mold used to form the two halves of the lamp, Fig. 2.

Molding the Hanging Lamp

1. Apply a coat of mold release wax to the metal surface of the mold.
2. Make the bottom half of the lamp using two layers of glass mat saturated with polyester resin. Color the resin to make the layup translucent, yet not allow glare from the lamp.
3. Cover the layup with Saranwrap and roll the air bubbles out by use of a brayer.
4. Remove the Saranwrap before the layup is cured. This will leave a slightly raised fiber pattern texture. Pat any separated fibers back to the surface with a brush. Remove the laminate from the mold after it has hardened.
5. For the top half of the lamp, use two layers of mat and clear resin with a piece of perforated aluminum between the layers for added decoration. After the laminate has hardened, remove it from the mold.
6. Lay out the two laminated halves, saw them to shape, sand and assemble to form the completed lamp, (Project 47).

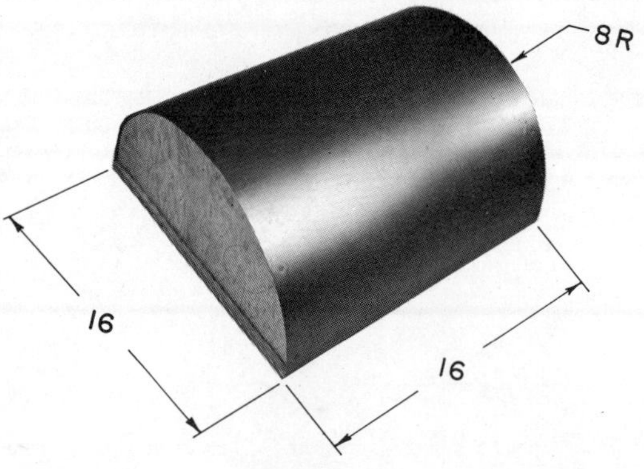

Fig. 1. Sheet Metal and Plywood Mold for Hanging Lamp

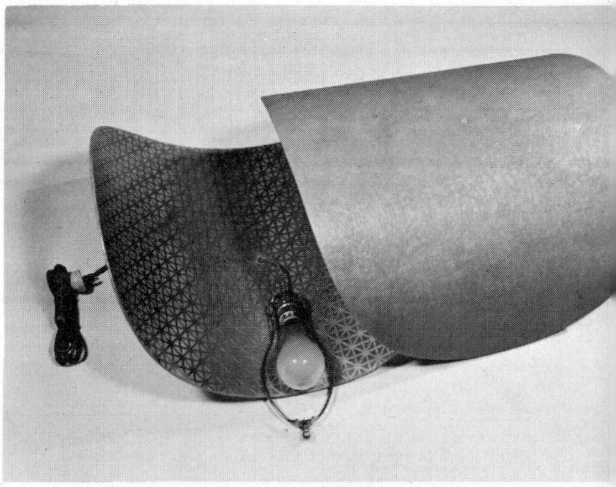

Fig. 2. Hanging Lamp Disassembled

TECHNIQUE 10

Glass Fiber Reinforced Plastics Mold Making

To make a glass fiber reinforced plastics mold, it is necessary to first make a wood mold, Fig. 1, by employing patternmaking practices. Internal fillets are made with fillet wax and the completed mold is then coated with epoxy resin (100 percent solids). When the resin has dried, wet-sand the mold with number 400 grit wet-or-dry sandpaper. Then coat it with release wax and polyvinyl alcohol, allowing each coat to dry before the succeeding one is applied. The wood mold is now ready to be used for making a glass fiber reinforced plastics mold.

Making the Mold

1. Apply a gel coat of polyester resin to the wood mold. A gel coat is a thin outer layer of resin applied to the mold. Follow the manufacturer's specifications for mixing the proper amount of hardener (catalyst) with the resin.

2. Allow the gel coat to "set" so that the fibers of the glass mat will not penetrate it when applied.

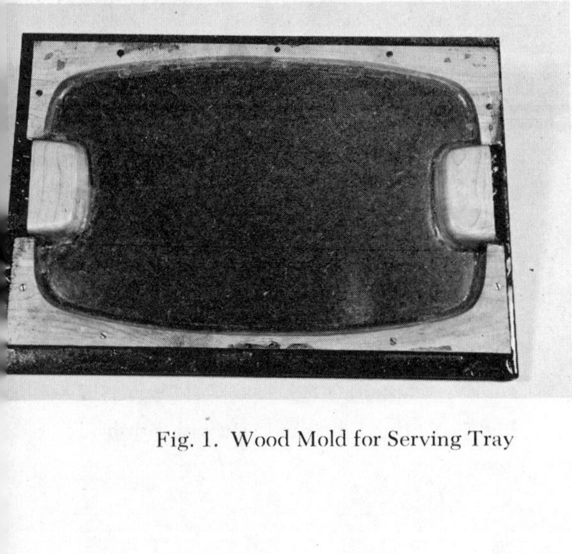

Fig. 1. Wood Mold for Serving Tray

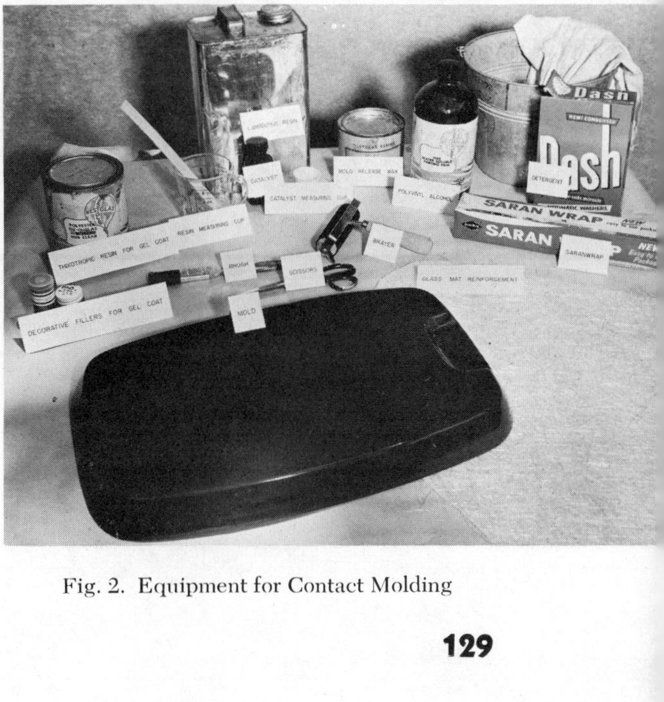

Fig. 2. Equipment for Contact Molding

3. Place a layer of glass mat on the wood mold and saturate it with resin.

4. Cut a piece of plywood to fit on top and around the handles of the wood mold.

5. Place this plywood on top of the layup, draw the edges of the layup over the top of the plywood and tack them in place.

6. Cut a second piece of plywood to fit over the first piece and fasten it in

Fig. 3. Applying Gel Coat to Mold

Fig. 4. Applying Resin and Glass Mat

Fig. 5. Rolling Saranwrap-Covered Layup to Remove Air

Fig. 6. Removing Saranwrap

Fig. 7. Patting Fibers in Place with Brush

place with screws, thereby holding the layup edges in place.

7. Clamp the entire assembly until cure is completed.

8. After the glass fiber reinforced plastics mold has cured, remove it from the wood mold and apply a gel coat to the exposed plywood edges.

9. Wet-sand the entire mold, using successively finer grades of abrasive paper, until the surface is smooth. The completed mold is shown in Fig. 2. This type of mold is widely employed in industry, and may be used indefinitely as long as the release agents are renewed for each molding.

Molding the Serving Tray

1. Apply a coat of wax, two coats of polyvinyl alcohol, and a final wax coat to the mold to form the release agent.

2. Mix a gel coat containing bronze powder and silver "glitter" and apply it over the release agent as shown in Fig. 3.

3. Apply two layers of glass mat, saturating each with catalyzed resin by dabbing with a brush, Fig. 4.

4. Place Saranwrap over the flat bottom of the layup and roll out air bubbles with a brayer, Fig. 5.

5. Peel off the Saranwrap and dab down the fibers over the entire surface including the edges, Figs. 6 and 7.

6. Allow the layup to cure, then remove the hardened laminate, Fig. 8, by using compressed air around the edges in addition to prying gently with the fingers.

7. Mark the laminate for trimming as shown in Fig. 9.

8. Trim the marked laminate on the band saw, then sand it to smooth the edges, Figs. 10, 11, and 12. Fig. 13 shows the completed tray.

Fig. 9. Marking the Trim Line

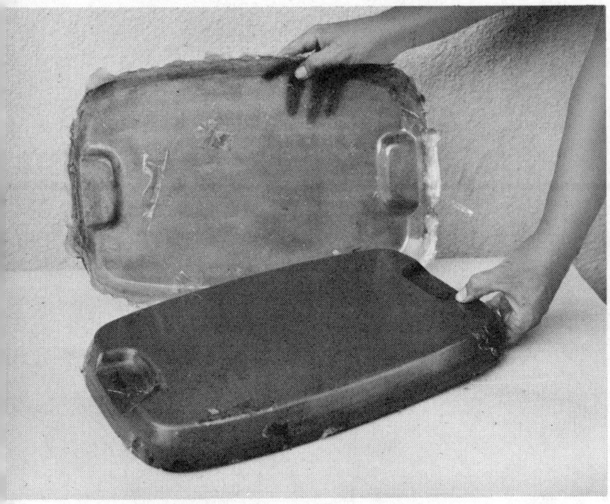

Fig. 8. Laminate Removed from Form

Fig. 10. Trimming Laminate on Band Saw

Fig. 11. Sanding Edges on Belt Sander

Fig. 12. Final Sanding to Smooth Edge

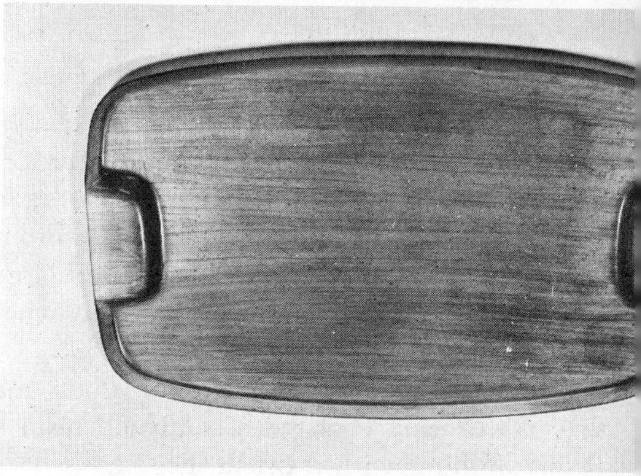

Fig. 13. Completed Serving Tray

Bibliography

Selected on the basis of appropriateness for the student and hobbyist, these books will provide a background which will prove helpful in the use of this book.

Adams, John V., *Plastic Arts Crafts*. New York: D. Van Nostrand Company, Inc., 1949, i-vii, 147 p.

This is a book for the beginner in the art of plastic crafts, giving step-by-step procedures for planning and making plastic articles. It is well illustrated with action-type photographs showing proper techniques. It also includes a brief history and classification of plastics, and a description of the types and characteristics of various plastics that can be used in craft work.

Butzko, Robert L., *Plastic Sheet Forming*. New York: Reinhold Publishing Corp., 1958, 178 p.

This book describes air pressure and vacuum forming. Also included are specific processes and methods of selecting proper materials. Various molds and their problems in design are discussed. Commercial sheet forming processes are included.

Cherry, Raymond, *General Plastics: Projects and Procedures.* Bloomington, Illinois: McKnight & McKnight Publishing Company, 1948, 156 p.

Contained in this book are detailed illustrations and descriptions of hand tool and machine operations, and complete instructions for making projects from cast phenolics, acrylics, and acetates. Also included is a brief history of plastics; information on their composition, common trade names, and various applications. The use of findings in finishing certain projects and a list of references are included.

Cope, Dwight, *Cope's Plastic Book.* Chicago: The Goodheart-Willcox Co., Inc., 1960, 272 p.

Over ninety plastic projects are illustrated, ranging from elementary to advanced. A description of various types of plastic and how they are made is included. It also describes methods of working with plastics – storing, sawing, machining, polishing, heat forming, cementing, bonding, cleaning, internal carving, dyeing, and commercial fabrication methods.

DeWick, Ernest S., *Plastic Craft.* New York: The Macmillan Company, 1949, i-xii, 184 p.

A high school approach to plastics is presented by layouts of projects, lists and illustrations of the use of tools, and visual applications.

Dunham, Arthur, *Working With Plastics.* New York: McGraw-Hill Book Company, Inc., 1948, i-x, 324 p.

This book deals primarily with cast phenolic and acrylic plastics. History and composition of plastics is discussed. A number of projects are presented and described, including various methods and techniques of working plastics.

Groneman, Chris H., *Plastics Made Practical.* Milwaukee: Bruce Publishing Co., 1948, i-x, 324 p.

Included are history and development; tools and methods, and suggested projects. All procedures are well illustrated.

Ickis, Marguerite, *Working in Plastics.* New York: Barmaroy Publishing Co., 1947, 48 p.

This is a short, completely illustrated book that covers standard forms, tools, and procedures for working plastic. With each different procedure, several projects are shown.

Lockrey, A. J., *Plastics in the School and Home Workshop.* New York: D. Van Nostrand Company, Inc., 1946, i-vi, 235 p.

This book contains general information on working plastics, related commercial processes, and project construction. It is well illustrated and contains drawings of all projects.

Mansperger, Dale E. and Pepper, Carson W., *Plastics Problems and Processes.* Scranton: International Textbook Company, 1942, i-xxii, 350 p.

This book presents the history of plastics; also practical applications and manufacturing processes.

Newkirk, Louis V., *Adventures With Plastics.* New York: D. C. Heath and Company, 1947, i-xii, 275 p.

This book was written for the person who is just beginning to work with plastics in the school or home workshop. Common tools and machines used in working with plastics are discussed; project designs and construction suggestions are illustrated. Information concerning various types of plastics and the common industrial processes used in the manufacture of plastic items, as well as how to order plastics and craft supplies are also included.

Plastic Catalog Corp., *Modern Plastics Encyclopedia.* Bristol, Connecticut: Breskin Publications, Inc., 1963, 1195 p. (Published Annually)

This book of the plastic industrial scene presents current technical information for industrial application of plastics, and basic information for the non-technical user of plastics. Included are: a guide to plastics with property charts for all plastics; sections on each use of plastic (forms, molding, reinforcing, laminating, and finishing). Related information includes chemicals and adhesives, fillers, engineering methods, and equipment. It is a resource book listing major manufacturers of plastics and their industrial application.

Robinson, Clark N., *Meet the Plastics*. New York: The Macmillan Co., 1949, i-x, 168 p.

The approach used is to provide consumers of plastics with general information, but it is technical enough to present specialization possibilities for future technicians of the plastic industry. A directory lists manufacturers, publications, bibliography, and training facilities.

Rohm and Haas Company, *Plexiglas Craftsman's Handbook*. Philadelphia: Rohm and Haas, 1955.

Written primarily for the novice in acrylic plastic craftwork, projects are presented which range in skills from the very simple up to advanced. Each project lists the requirements in tools, supplies, and materials. If followed step-by-step, these projects provide a logical, easy and interesting training program. The book also explains and illustrates various operations, such as machining, forming, cementing, polishing, internal carving, handling, and storing.

Rohm and Haas Company, *Plexiglas Design and Fabrication Manual*. Philadelphia: Rohm and Haas, 1955.

Contains technical information on cutting, machining, forming, cementing, finishing, annealing, and designing with acrylic plastic. Also gives much general information needed and used by plastic fabricators including many illustrations.

Steele, Gerald L., *Fiber Glass: Projects and Procedures*. Bloomington, Illinois. McKnight & McKnight Publishing Company, 1962, 159 p.

This book specializes in fiber glass projects and their construction in the school or home. Includes characteristics of materials and the equipment and processes used in working with fiber glass, as well as a glossary and safety sections. Information on reinforcement characteristics may be found in the appendix, which also includes lists of manufacturers of fiber glass and supply sources.

Stoddard, H. W., *Constructional Plastics and Allied Crafts*. London: J. M. Dent and Sons, Ltd., 1951, i-viii, 108 p.

A course in plastics and related plastic crafts, including plastic and bead jewelry crafts. Includes tools, equipment, and techniques used in working with plastics on the industrial arts level with many illustrations and project ideas.

Wolfe, Barnard, *Plastics: What Everyone Should Know*. Indianapolis: The Bobbs-Merrill Company, 1945, i-vi, 189 p.

This book gives a general picture of plastics to the layman, including the development of many plastics and their uses in products in an easily understood form.

Index